I0467998

NUREG–1472

Regulatory Analysis for the Resolution of Generic Issue 57: Effects of Fire Protection System Actuation on Safety-Related Equipment

Manuscript Completed: October 1993
Date Published: October 1993

H. W. Woods

**Division of Safety Issue Resolution
Office of Nuclear Regulatory Research
U.S. Nuclear Regulatory Commission
Washington, DC 20555–0001**

ABSTRACT

Actuation of Fire Protection Systems (FPS) in Nuclear Power Plants have resulted in adverse interactions with equipment important to safety. Precursor operational experience has shown that 37% of all FPS actuations damaged some equipment, and 20% of all FPS actuations have resulted in a plant transient and reactor trip. On an average, 0.17 FPS actuations per reactor year have been experienced in nuclear power plants in this country. This report presents the regulatory analysis for GI–57, "Effects of Fire Protection System Actuation on Safety-Related Equipment". The risk reduction estimates, cost/benefit analyses, and other insights gained during this effort have shown that implementation of the recommendations contained in this report can significantly reduce risk, and that these improvements can be warranted in accordance with the backfit rule, 10 CFR 50.109(a)(3). However, plant specific analyses are required in order to identify such improvements. Generic analyses can not serve to identify improvements that could be warranted for individual, specific plants. Plant specific analyses of the type needed for this purpose are underway as part of the Individual Plant Examination of External Events (IPEEE) program.

CONTENTS

Page

Abstract iii

List of Tables vii

Executive Summary ix

1 Statement Of The Problem 1

2 Objective 2

3 Alternative Resolutions 2

   3.1 Alternative 1 – No Action 2

   3.2 Alternative 2 – Upgrade Relay-Based FPS Actuation Controllers with
       Seismically Qualified Printed Circuit Boards 2

   3.3 Alternative 3 – Upgrade Fire Protection Systems from Smoke Detector
       Actuated to Heat Detector Actuated 2

   3.4 Alternative 4 – Reroute Safety-Related Cables 2

   3.5 Alternative 5 – Seismically Qualify CO_2 Tanks, Outlet Piping, and Battery Rack 3

   3.6 Alternative 6 – Seismically Qualify FPS Battery Racks 3

   3.7 Alternative 7 – Upgrade the FPS Water Quality 3

   3.8 Alternative 8 – Replace Deluge with Preaction Sprinkler FPS in Selected Plant Locations 3

   3.9 Alternative 9 – Upgrade Electrical Cabinet Design to Prevent Water Intrusion 3

   3.10 Alternative 10 – Seismically Anchor Safety-Related Cabinets Susceptible
        to Tipping/Sliding Failures 3

4 Technical Findings 3

   4.1 Core Damage Frequency Analysis 4

      4.1.1 Generic FPS Root-Cause Actuation Scenarios 4

      4.1.2 Generic Plant Analysis 5

      4.1.3 Generic Core Damage Frequency Estimates 7

   4.2 Dose Consequences Analysis 10

   4.3 Cost Analysis 11

      4.3.1 Labor and Equipment/Materials Costs 11

      4.3.2 Engineering and Quality Assurance/Control Costs 12

      4.3.3 Radiation Exposure 12

      4.3.4 Health Physics Support Costs 12

      4.3.5 Anti-Contamination Clothing Costs 13

      4.3.6 Radioactive Waste Disposal Cost 13

      4.3.7 Other Licensee Costs 13

      4.3.8 NRC Costs 13

      4.3.9 Onsite Averted Costs 13

   4.4 Cost Estimate Uncertainties 14

   4.5 Backfit Alternatives Cost Estimates 14

      4.5.1 Alternative 1 – No Action 14

      4.5.2 Alternative 2 – Upgrade a FPS Actuation Controller with Seismically Qualified
          Printed Circuit Boards 14

      4.5.3 Alternative 3 – Replace Smoke Detector Actuated FPS with Heat Detector
          Actuated System 14

      4.5.4 Alternative 4 – Reroute Safety-Related Cables 14

      4.5.5 Alternative 5 – Seismically Qualify the CO_2 Tank, Outlet Piping, and Battery Rack 15

CONTENTS (Continued)

			Page
	4.5.6	Alternative 6 – Seismically Qualify A FPS Battery Rack	15
	4.5.7	Alternative 7 – Upgrade the FPS Water Quality	15
	4.5.8	Alternative 8 – Replace Deluge with Preaction Sprinkler FPS	15
	4.5.9	Alternative 9 – Replace Electrical Cabinet with a Cabinet Designed to Prevent Water Intrusion	16
	4.5.10	Alternative 10 – Seismically Anchor Safety-Related Cabinets Susceptible to Tipping/Sliding Failure	16
5	**Value/Impact Analysis**		**17**
	5.1	Backfit Analysis	17
	5.1.1	Alternative 1 – No Action	17
	5.1.2	Alternative 2 – Upgrade an FPS Actuation Controller with Seismically Qualified Printed Circuit Boards	17
	5.1.3	Alternative 3 – Replace Smoke Detector Actuated FPS with a Heat Detector Actuated System	17
	5.1.4	Alternative 4 – Reroute Safety-Related Cables	17
	5.1.5	Alternative 5 – Seismically Qualify the CO_2 Tank, Outlet Piping, and Battery Rack	17
	5.1.6	Alternative 6 – Seismically Qualify A FPS Battery Rack	18
	5.1.7	Alternative 7 – Upgrade the FPS Water Quality	18
	5.1.8	Alternative 8 – Replace Deluge with Preaction Sprinkler FPS	20
	5.1.9	Alternative 9 – Replace Electrical Cabinet with a Cabinet Designed to Prevent Water Intrusion	20
	5.1.10	Alternative 10 – Seismically Anchor Safety-Related Cabinets Susceptible to Tipping/Sliding Failure	22
	5.2	Frontfit Analysis	22
	5.2.1	Alternative 1 – No Action	23
	5.2.2	Alternative 2 – Upgrade a FPS Actuation Controller with Seismically Qualified Printed Circuit Boards	23
	5.2.3	Alternative 3 – Replace Smoke Detector Actuated FPS with a Heat Detector Actuated FPS	23
	5.2.4	Alternative 4 – Reroute safety-related Cables	24
	5.2.5	Alternative 5 – Seismically Qualify the CO_2 Tank, Outlet Piping, and Battery Rack	24
	5.2.6	Alternative 6 – Seismically Qualify a FPS Battery Rack	24
	5.2.7	Alternative 7 – Upgrade the FPS Water Quality	24
	5.2.8	Alternative 8 – Replace Deluge with Preaction Sprinkler FPS	24
	5.2.9	Alternative 9 – Replace Electrical Cabinet with a Cabinet Designed to Prevent Water Intrusion	24
	5.2.10	Alternative 10 – Seismically Anchor Safety-Related Cabinets Susceptible to Tipping/Sliding Failure	24
6	**Decision Rationale**		**25**
7	**Implementation**		**27**
8	**References**		**27**

LIST OF TABLES

Page

4.1	Fire Protection System Summary	6
4.2	Fire Protection Cases for Analysis	6
4.3	Fire Occurrence Frequencies	6
4.4	FPS Actuation Frequency per System-Year	7
4.5	Core Damage Frequency – "Most Vulnerable" Generic Plant	8
4.6	Core Damage Frequency – "Typical" Generic Plant	9
4.7	Core Damage Frequency – "Least Vulnerable" Generic Plant	9
4.8	Plant-Specific Base Case Results in Terms of Core Damage Frequency (Per Reactor Year)	10
4.9	Risk in person-Sv/RY (person-rem/RY) "Typical" PWR Plant	11
4.10	Risk in person-Sv/RY (person-rem/RY) "Typical" BWR Plant	12
5.1	Alternative 2 $K/person-Sv ($K/person-rem) averted	18
5.2	Cost/Benefit Ratio Alternative 3 $K/person-SV ($K/person-rem) averted	19
5.3	Cost/Benefit Ratio Alternative 4 $K/person-Sv ($K/person-rem) averted	19
5.4	Cost/Benefit Ratio Alternative 5 $K/person-Sv ($K/person-rem) averted	20
5.5	Cost/Benefit Ratio Alternative 6 $K/person-Sv ($K/person-rem) averted	21
5.6	Cost/Benefit Ratio Alternative 8 $K/person-Sv ($K/person-rem) averted	21
5.7	Cost/Benefit Ratio Alternative 9 $K/person-Sv ($K/person-rem) averted	22
5.8	Cost/Benefit Ratio Alternative 10 $K/person-Sv ($K/person-rem) averted	23

EXECUTIVE SUMMARY

This report provides supporting information, including a value/impact analysis for the Nuclear Regulatory Commission's (NRC's) resolution of Generic Issue 57, "Effects of Fire Protection System Actuation on Safety-Related Equipment". This issue addresses the concerns regarding the actuation of fire protection systems (FPS) using water, carbon dioxide and halon, and the effects of these fire suppressants on plant equipment, the damage or upset of which might initiate and sustain risk significant event sequences.

FPS actuations that result in adverse interactions with plant systems needed to achieve safe plant shutdown or to mitigate a postulated accident reduce the availability of such systems. This concern is accentuated when common cause initiators and common mode failures of safety-related equipment are considered. Examples of common cause initiators include earthquakes, smoke and heat intrusion into multiple fire zones, and fire suppressant intrusion into multiple fire zones affecting several safety-related systems. Examples of common mode failures of safety-related systems and/or auxiliary systems supporting safety-related systems include electrical shorts in instrument cabinets and electrical power distribution centers, CO_2 ingress into the fresh air intake of emergency diesel-generator sets, and cold CO_2 induced thermal stresses and cracking of station battery casings, with loss of offsite power during an earthquake. It should be noted that a number of common cause initiators and common mode failures are not mutually exclusive and they may be part of a single event sequence.

Generic Issue 57 was identified in 1982 (Refs. 1,2) as a result of a number of precursor events showing that safety-related equipment subjected to fire protection system (FPS) water spray could be rendered inoperable. These precursor events also indicated numerous spurious FPS actuations initiated by operator testing errors or by maintenance activities, e.g., welding, and steam or high humidity in the vicinity of FPS detectors or control circuitry and components. An NRC memorandum (Ref.2), issued on January 28, 1982, provided additional examples of FPS actuation interactions and suggested that all types of FPS be included in a review of their safety significance and possible corrective steps.

On June 22, 1983, IE Information Notice 83–41 (Ref.3) was issued to alert licensees and to provide examples of recent experiences in which actuation of fire protection systems caused damage or inoperability of safety-related systems. The effects of such events range from reactor trips to fires in high voltage electrical equipment, and water contamination of diesel fuel. The IE Information Notice indicated that the plant Fire Hazards Analysis under Appendix R to 10 CFR 50 requires, not only consideration of the consequences of a postulated fire, but also consideration of the effects of fire-fighting activities. The IE Information Notice also stated that a properly conducted Fire Hazards Analysis in conjunction with a physical walkdown of plant areas would have identified instances where minor modifications such as shielding equipment and sealing conduit ends would have reduced equipment water damage from inadvertent FPS operation.

Four plants representative of the various designs of currently operating nuclear power plants were evaluated as part of this issue (Refs. 4–8). Furthermore, a generic evaluation of this issue (Ref. 9) was performed taking into account the insights from the technical findings of these four evaluations, as well as design and plant layout information of a large number of operating plants collected for this purpose. An extensive review of operational experience involving actuations of fire protection systems was performed prior to the analytical assessments of risk associated with this issue. The review of the operational experience showed the following:

- 0.15 inadvertent FPS actuations/RY

- 0.02 advertent FPS actuations/RY

- 37% of all actuations damaged some equipment

- 20% of all actuations resulted in a plant transient and reactor trip

- Core Damage Frequency (CDF) contributions from GI–57 root causes for the four individual plants evaluated were estimated to be in the range of 7.3E–06/RY to 5.6E–05/RY.

- Dominant risk contributors are associated with seismic/FPS and seismic/fire interactions resulting in sequences involving station blackout and small LOCAs

Both of these categories of dominant sequences are currently being addressed by the Individual Plant Examination of External Events (IPEEE) program. Generic Letter 88–20, Supplement 4, June 27, 1991 (which initiated the IPEEE) states:

"The walkdown procedures should be specifically tailored to assess the remaining issues identified in the Fire Risk Scoping Study: (1) seismic/fire interactions, (2) effects of fire suppressants on safety equipment, ..." (paragraph 4.2).

The staff notes that any complete assessment of seismic/fire interactions would necessarily include seismic effects on manual firefighting, since automatic fire suppressant systems are not seismically qualified, thus increasing the potential need for effective manual firefighting during seismic events.

NUREG–1407 ("Procedural and Submittal Guidance for the IPEEE for Severe Accident Vulnerabilities", June, 1991) reiterates the above (in the first paragraph of section 4) and also adds:

> "The use of an existing fire PRA for the internal fires IPEEE is acceptable provided the PRA reflects the current as-built and as-operated status of the plant and the licensee addresses the deficiencies of past PRAs that are identified in the Fire Risk Scoping Study (NUREG/CR–5088). Deficiencies may include the use of low conditional failure probabilities for dampers and penetrations, no consideration of damage from the use of fire suppressants, inappropriate estimates of the effectiveness of manual fire fighting, and no consideration of seismic/fire interactions." (paragraph 4.2).

Also in NUREG–1407, Appendix D, section 6 ("Internal Fires") the staff response to question 6.2 states:

> "The procedurally directed walk-downs associated with internal fires vulnerability evaluation can be planned as part of the seismic walk-downs that would specifically look for the seismic-induced fire vulnerability issues. The idea is to first identify those areas that could be vulnerable so that they can be brought into focus during the walkdown.

> "For example, if a plant didn't have its diesel fuel tank strapped down properly one could postulate a large fuel source for fire as a result of a seismic event. Other similar seismic/fire interactions were summarized in Section 7 of NUREG/CR–5088."

In addition, for performance of the IPEEE, many licensees will also be using the Fire-Induced Vulnerability Evaluation (FIVE) methodology developed by the Electric Power Research Institute (EPRI) as described in "Fire-Induced Vulnerability Evaluation (FIVE)", EPRI TR–100370, April, 1992. Following a description of the three basic Phases of FIVE, that document states:

> In addition, there is a discussion of several potentially risk significant items that were identified in the NUREG/CR–5088, "Fire Risk Scoping Study", that **should also be considered in performing FIVE**. (Section 7.0) **[emphasis** added].

Section 7.0 of NUREG/CR–5088 is a description of the Sandia Fire Risk Scoping Study Evaluation which includes discussion of the dominant risk contributors to GSI–57-related events, including Seismic/Fire interactions (seismically induced fires, seismic actuation of fire suppression systems, and seismic degradation of fire suppressant systems), and Manual Fire Fighting Effectiveness. The following seismic/fire interactions are given as examples of situations to be considered by the IPEEE: unanchored CO_2 or Halon tanks, possible relay chatter in fire protection system actuation systems, fire alarm systems having only a smoke-actuated alarm without a heat or flame detector, fire pump mounts without vibration amplitude stops, cast iron fire mains, inadequate anchoring of electrical cabinets and inadequate slack in the wires leading to such cabinets (to avoid sparks from tight wires), unanchored high-pressure gas bottles, potential interactions between sprinkler system heads and adjacent pipes, and presence of mercury switches in fire suppression and detection systems (such switches should be replaced with alternate, seismically insensitive switches).

For the "generic" evaluation, after subtraction of CDF contributions from GSI–57-related events involving:

(a) <u>Seismic/Fire</u>: – seismic induced fire plus seismic induced suppressant diversion. The unsuppressed fire and/or the diverted suppressant incapacitate safety related equipment needed to mitigate effects of the seismic event; and

(b) <u>Seismic/FPS</u>: – seismic induced actuation of the FPS. Released suppressant damages safety-related equipment needed to mitigate effects of the seismic event,

which are being emphasized by the IPEEE, the mean CDF of the remaining contributors is less than 1.0E–05/RY which does not justify a generic backfit.

The risk reduction estimates, cost/benefit analyses, and other insights gained during this effort have shown that implementation of the recommendations contained in this report can significantly reduce risk, and that these improvements can be warranted in accordance with the backfit rule, 10 CFR 50.109(a)(3). However, plant specific analyses are required in order to identify such improvements. Generic analyses can not serve to identify improvements that could be warranted for individual, specific plants. Plant specific analyses of the type needed for this purpose are underway as part of the Individual Plant Examination of External Events (IPEEE) program.

Hence, we concluded that GSI–57 should be resolved with no new regulatory requirements generically imposed for existing plants.

For new plants, the enhanced safety requirements given in SECY–90–016, "Evolutionary Light Water Reactor (LWR) Certification Issues and their Relationship to Current Regulatory Requirements", January 12, 1990, allow the same GSI–57 resolution to be applied (i.e., with no new regulatory requirements imposed). This is justified since the designers of new plants will perform plant-specific PRAs, including external events, that are equivalent to the IPEs and IPEEEs which will resolve this issue for existing plants.

However, the reports dealing with the evaluation of this issue, including this regulatory analysis, will be published, or otherwise be made publicly available, so that the insights gained from this effort may be used by licensees and applicants, and the nuclear industry in general, to take voluntary steps for plant-specific cases, including frontfitting considerations for advanced or evolutionary reactor designs.

1 STATEMENT OF THE PROBLEM

Generic Issue 57, "Effects of Fire Protection System Actuation on Safety-Related Equipment", was identified in 1982 (Refs. 1,2) as a result of a number of precursor events showing that safety-related equipment subjected to fire protection system (FPS) water spray could be rendered inoperable. These precursor events also indicated numerous spurious FPS actuations initiated by operator testing errors or by maintenance activities, e.g., welding, and steam or high humidity in the vicinity of FPS detectors or control circuitry and components. An NRC memorandum (Ref.2), issued on January 28, 1982, provided additional examples of FPS actuation interactions and suggested that all types of FPS, e.g., water, halon, carbon dioxide and other chemicals be included in a review of their safety significance and possible corrective steps.

A review of FPS related regulations and guidelines regarding interactions between FPS features and plant safety systems, as well as a review of operating experience involving such interactions, led to the conclusion that, if existing requirements are properly implemented, such interactions should be minimized. GDC–3 states that inadvertent FPS operation or failure should not impair the function of safety systems. However, licensee event reports (LERs) indicated that regulatory requirements, not necessarily limited to GDC–3, had not been properly implemented at some plants. For example, deficiencies in sealing of electrical cable conduits resulted in the transport of water in instrument and electrical cabinets resulting in equipment damage or upset including FPS actuations.

On June 22, 1983, IE Information Notice 83–41 (Ref.3) was issued to alert licensees and to provide examples of recent experiences in which actuation of fire protection systems caused damage or inoperability of safety-related systems. The effects of such events range from reactor trips to fires in high voltage electrical equipment, and water contamination of diesel fuel. The IE Information Notice indicated that the plant Fire Hazards Analysis under Appendix R to 10 CFR 50 requires, not only consideration of the consequences of a postulated fire, but also consideration of the effects of fire-fighting activities. The IE Information Notice also stated that a properly conducted Fire Hazards Analysis in conjunction with a physical walkdown of plant areas would have identified instances where minor modifications such as shielding equipment and sealing conduit ends would have reduced equipment water damage from inadvertent FPS operation. The IE Information Notice indicated that none of the reported events resulted in a serious impact on the functional capability of a plant to protect public health and safety. However, examples were given where it would

not be difficult to extrapolate actual occurrences into a sequence of events that could lead to more serious consequences.

FPS actuations that result in adverse interactions with plant systems needed to achieve safe plant shutdown or to mitigate a postulated accident reduce the availability of such systems. This concern is accentuated when common cause initiators and common mode failures of safety-related equipment are considered. Examples of common cause initiators include earthquakes, smoke and heat intrusion into multiple fire zones, and fire suppressant intrusion into multiple fire zones affecting several safety-related systems. Examples of common mode failures of safety-related systems and/or auxiliary systems supporting safety-related systems include electrical shorts in instrument cabinets and electrical power distribution centers, CO_2 ingress into the fresh air intake of emergency diesel-generator sets, and cold CO_2 induced thermal stresses and cracking of station battery casings, with loss of offsite power during an earthquake. It should be noted that a number of common cause initiators and common mode failures are not mutually exclusive and they may be part of a single event sequence.

Four plants representative of the various designs of currently operating nuclear power plants were evaluated as part of this issue (Refs. 4–8). Furthermore, a generic evaluation of this issue (Ref. 9) was performed taking into account the insights from the technical findings of these four evaluations, as well as design and plant layout information of a large number of operating plants collected for this purpose. An extensive review of operational experience involving actuations of fire protection systems was performed prior to the analytical assessments of risk associated with this issue. The review of the operational experience showed the following:

- 0.15 inadvertent FPS actuations/RY

- 0.02 advertent FPS actuations/RY

- 37% of all actuations damaged some equipment

- 20% of all actuations resulted in a plant transient and reactor trip

- Core Damage Frequency (CDF) contributions from GI–57 root causes for the four individual plants evaluated were estimated to be in the range of 7.3E–06/RY to 5.6E–05/RY.

- Dominant risk contributors are associated with seismic/FPS and seismic/fire interactions resulting in sequences involving station blackout and small LOCAs

2 OBJECTIVE

The objective of the Generic Issue 57 resolution program is to evaluate the present design of fire protection systems in operating nuclear power plants and the effects of their actuation on plant safety, and to examine the cost effectiveness of alternative measures for reducing the overall vulnerability of safety systems to FPS actuations. It is also the objective of this issue to assess these alternative corrective measures for their applicability and efficacy to evolutionary or new designs of nuclear power plants.

Probabilistic methods were used to assess the CDF, the potential reduction in risk of alternative corrective measures and their cost-effectiveness. The overall objective of the resolution of GI-57 is that contribution from FPS actuations should be a small percentage of the total CDF from all causes.

For USI A-45, the staff recommended that the frequency of events related to decay heat removal failure leading to core damage should be reduced to a level (about 1.0E-05/RY) that the probability of such an accident in the next 30 years would be about 0.03 based on a population of about 110 plants. A CDF objective of 1.0E-05/RY was also noted in USI A-44, "Station Blackout." These objectives are also consistent with the recently issued guidance to the staff (Ref. 10) setting a safety goal of less than 1.0E-04/RY for CDF from all contributors, with a subsidiary safety goal of no more than 1.0E-05/RY for a single contributor. The interim guidance provided in Reference 10 for purposes of satisfying the substantial additional protection criterion of the backfit rule sets forth the guideline that initiatives involving actions to reduce CDF should result in at least 1.0E-05/RY reduction. In assessing the risk reduction significance of potential backfits the conditional containment failure probability (CCFP) must also be evaluated against the guidance provided in Reference 10. Section 6 of this regulatory analysis takes into account all these considerations, including attendant uncertainty and sensitivity analyses in developing the decision rationale for resolving this issue.

3 ALTERNATIVE RESOLUTIONS

There were several alternatives analyzed for the resolution of Generic Issue 57. These alternatives are described below:

3.1 Alternative 1 – No Action

With this alternative, there would be no regulatory requirements. Consistent with existing regulations, this alternative does not preclude a licensee, or an applicant for an operating license, from proposing to the NRC staff design changes intended to enhance the reliability or operability of the protection systems and their components on a plant-specific basis.

3.2 Alternative 2 – Upgrade Relay-Based FPS Actuation Controllers with Seismically Qualified Printed Circuit Boards

During our evaluation of GI-57 it was determined that many types of relays are used for the actuation of fire protection systems. In some plants, mercury wetted relays for fire protection system actuation, alarm annunciation and isolation of equipment room cooling are used. In the event of an earthquake it is highly likely that some of these relay types, and certainly the mercury-wetted ones would actuate. Under this alternative, relay-based controllers would be replaced with printed circuit boards to reduce the potential for inadvertent actuation of fire protection systems during earthquakes.

3.3 Alternative 3 – Upgrade Fire Protection Systems from Smoke Detector Actuated to Heat Detector Actuated

Smoke detector actuated systems are subject to more frequent inadvertent actuations. Modifying such systems to heat detector actuated systems would result in reducing the frequency of inadvertent FPS actuations. It might be prudent to retain the existing smoke detectors for indication or alarming only, or as a part of a coincident logic actuation system based on heat and smoke detectors.

3.4 Alternative 4 – Reroute Safety-Related Cables

Our evaluation of this issue, including some plant walkdowns, found that there are certain "pinch points" in plants where cables for some redundant safety-related systems are routed close together. Given a fire or FPS actuation which would damage these cables, the respective safety-related systems would be vulnerable to common failure. Hence, the purpose of this alternative would be to reroute one of the sets of cables to remove this common mode failure vulnerability.

3.5 Alternative 5 – Seismically Qualify CO_2 Tanks, Outlet Piping, and Battery Rack

A seismic walkdown and subsequent fragility analysis for a typical CO_2 system found a high likelihood of suppressant agent diversion given an earthquake. Failure of the tank or its outlet piping or the FPS battery dominated the overall probability of failure of the CO_2 system. This potential plant modification would seismically qualify the CO_2 tank, battery, and its immediate outlet piping.

3.6 Alternative 6 – Seismically Qualify FPS Battery Racks

It was found during a plant specific walkdown, that a water based FPS for an entire plant had two electric driven fire pumps and one diesel driven fire pump. Given a LOSP, both electric driven pumps would fail due to being powered from non-vital busses. A failure of the starting battery for the diesel pump would lead to a loss of fire main pressure.

3.7 Alternative 7 – Upgrade the FPS Water Quality

Water is the most frequently used FPS agent at nuclear power plants. The most common delivery systems are pre-action sprinklers, deluge wet-pipe sprinklers, dry-pipe sprinklers and standpipe hose systems. It has been postulated that improvement of water quality will reduce the potential for damage to safety-related components from exposure to water suppressant. To address this concern, a modification to upgrade the FPS water quality is performed. For this modification a water purification system would be required along with the associated piping and valves and a storage tank. It is assumed for this modification that the existing FPS piping and pumps would be utilized.

3.8 Alternative 8 – Replace Deluge with Preaction Sprinkler FPS in Selected Plant Locations

Of all of the types of water based FPSs utilized in nuclear power plants the pre-action sprinkler system has been found to experience the least number of inadvertent ac-

tuations (Appendix A). A pre-action sprinkler system requires the opening of a deluge value, either automatically, with a control signal, or manually, and a rise in the ambient temperature to the melting point of the fusible links on the sealed sprinkler heads. A deluge FPS only requires the opening of a valve, either automatically or manually to discharge water under pressure to the open spray heads. This modification involves replacing a deluge FPS with a pre-action sprinkler FPS. This modification would reduce the frequency of inadvertent actuations and localize the application of FPS agent.

3.9 Alternative 9 – Upgrade Electrical Cabinet Design to Prevent Water Intrusion

One study (Ref. 11) has reported that electrical equipment failure modes related to water were mainly due to electrical shorting and long term corrosion. Wherever water intrusion is possible the equipment could be expected to fail through shorting, grounding, tripping of overcurrent devices, physical damage due to the velocity of direct hose streams or long term corrosion causing potential failure of electro-mechanical parts. These failure modes were found to be dependent on the National Electrical Manufacturers Association (NEMA) rating or the configuration of the electrical enclosure. The appropriate NEMA rating to preclude water intrusion can potentially eliminate failures in electrical equipment due to water based FPSs.

3.10 Alternative 10 – Seismically Anchor Safety-Related Cabinets Susceptible to Tipping/Sliding Failures

In a seismic event energized cabinets present potential sources of fire. Although it is assumed that in a seismic event offsite power will be lost, thus deenergizing many electrical cabinets, there will be a number of safety-related cabinets energized by alternative power sources (batteries, diesel generators). These energized cabinets are susceptible to tipping/sliding failure possibly leading to a fire.

It is assumed that this modification would eliminate the potential for seismically induced fires due to the tipping/sliding failure of an energized electrical cabinet.

4 TECHNICAL FINDINGS

Sandia National Laboratories (SNL) was the primary contractor for performing the technical evaluation of the effects of fire protection system actuation on safety-

related equipment. This technical evaluation included a determination of the frequency of FPS actuations, core damage frequency from accident sequences resulting from FPS actuations, dose consequence analysis, and cost/benefit analysis. The detailed evaluations of specific plant designs are presented in References 4–8 and the generic, summary evaluation report is contained in Reference 9, which also contains an evaluation of the operational experience involving FPS actuations and the SNL methodology of performing the vital area and seismic analyses. This discussion of the technical findings is a condensed presentation drawn from the above references and Reference 9 in particular.

4.1 Core Damage Frequency Analysis

The core damage vulnerability caused by FPS actuations with the potential of causing damage to safety-related equipment was estimated using existing PRA models. A general transient event tree was used to account for FPS actuations. For seismically-induced FPS actuations, the loss of offsite power (LOSP) transient event tree was used. The success or failure of the required safety systems was determined by fault tree analyses identifying logically all possible combinations of component failures leading to the failure of the safety system in question. These logic models were combined using Boolean algebra as embodied in the SETS (Ref. 12) computer code, ultimately leading to the derivation of the "minimal cut sets" and accident sequences.

4.1.1 Generic FPS Root-Cause Actuation Scenarios

Based on the review of past experiences and walkdowns of a number of plants, thirteen generic root-cause scenarios, as shown in Table 3.1, were identified. Three root causes are due to inadvertent FPS actuations caused by a fire in another zone. Four are due to actuations resulting from purely random causes. Four are due to seismic causes, and one is due to the occurrence of a fire outside the plant. Also included is advertent FPS actuation with the presence of a fire. The various root causes of FPS actuation are described below and the specific tasks and information required to evaluate them are briefly discussed:

1. Fire-induced FPS actuation – FPS agent-induced damage. Based on the vital area analysis and plant specific (for example data submitted in accordance with 10CFR50 Appendix R), fire zones are identified where smoke or heat spread could cause inadvertent actuation in other plant areas which are either physically adjacent or connected through ventilation paths. Estimates are made of the impact of the FPS agent on equipment in these plant areas and are applied to the appropriate cut sets and accident sequences.

2. Fire-induced FPS actuation – recovery prevention. A plant's fire PRA is reviewed for risk-significant

recovery actions on equipment not damaged by fire and it is also determined in which plant areas these actions must occur. Then, other fire zones which are either physically adjacent or connected through ventilation are examined to determine if either heat or smoke spread could actuate the FPS and prevent the recovery action hypothesized. If any such combinations are found, the applicable accident sequences are requantified.

3. Fire-induced FPS actuation – access prevention. For each critical fire zone identified in the fire PRA where manual fire suppression was identified as the means of mitigating the fire, access to the fire zone is identified via plant specific data and a plant walkdown. As was the case for smoke or heat spread actuating a FPS and preventing recovery actions, a similar analysis is conducted for the delay in manual fire suppression caused by FPS actuation and the applicable accident sequences are requantified.

4. FPS actuation – human error. The vital area analysis and plant data are reviewed to determine which fire zones have an FPS that can be manually actuated. Erroneous manual actuation in any single fire zone may occur due to false detector signals or human errors of commission. For those cut sets requiring failures in more than one fire zone, the most likely scenario is that of a fire in one of the areas. Frequencies for such events are obtained from the historical data base for different types of FPSs. Using these frequencies, the accident sequences are requantified.

5. FPS actuation – steam pipe break. This root cause quantifies the core damage frequency contribution from inadvertent FPS actuation caused by a high-temperature steam environment. This actuation can occur due to moisture intrusion into a FPS controller, activation of smoke/heat detector(s), or melting of fusible link heads. An estimate of steam release frequency is made and the applicable accident sequences are requantified.

6. FPS actuation – hardware failures of FPS. In this scenario, inadvertent actuation of the FPS is caused by hardware failures of the FPS itself, such as a pipe break in a wet pipe system, or a failure in an FPS control circuit. Frequencies for such events are obtained from the historical data base for different types of FPSs. Using these frequencies, the accident sequences are requantified.

7. Seismic FPS actuation – dust. Those fire zones where automatic FPS are actuated solely by smoke/particulate detectors are identified. Then dust is assumed to cause FPS actuation in the fire zone given a seismic occurrence. The additional FPS damaged

equipment failures are added to the seismic sequences and these sequences are then requantified.

8. Seismic FPS actuation – relay chatter. The potential for seismically induced relay chatter is quantified based on a detailed evaluation of each FPS actuation circuit within a given plant. The additional FPS damaged equipment failures are added to the seismic sequences and these sequences are then requantified.

9. Seismic FPS actuation – mechanical failures. The potential for seismically induced mechanical failure is quantified based on a detailed evaluation and a plant walkdown of each FPS. The vital area equations can again be used directly to assess the impact of such events.

10. External fire-caused FPS actuation. Frequency of smoke intake from external fires is estimated from a combination of generic and plant-specific data. Fire zones potentially affected by smoke spread from outside ventilation are identified.

11. FPS actuation with the presence of a fire. Quantification of this scenario requires either an existing fire PRA or identification of fire sources in critical fire zones. Each fire zone with a FPS is identified to judge the effect of a fire with the simultaneous release of FPS agent on otherwise undamaged vital equipment. These failures are added to the fire sequences and requantified.

12. Seismic/fire interaction. In this scenario one or more seismically induced fires are evaluated for the probability of occurrence based on a plant walkdown and seismic fragility analysis of fire sources within the zone(s). The probability of diversion of FPS agents into zones not containing the fire(s) is made. These failures are added to the seismic sequences and requantified.

13. FPS actuation – unknown causes. In this scenario inadvertent actuation of the FPS is due to unknown causes. Frequencies for such events are obtained from the historical data base for different types of FPS. Using these frequencies, the accident sequences are requantified.

To identify the critical plant fire zones, criteria were developed for each root cause scenario which enable the analyst to determine which zones are potentially subject to each root cause of FPS actuation, given the general vital area analysis accident sequence equations. These criteria are shown in Table 3.2 of Reference 9. This step was performed manually, and required a review of plant systems, plant layouts, and plant specific data. This review considered such factors as the following:

- the presence of automatic or manual fixed FPSs,
- physical and electrical separation of redundant trains,
- susceptibility to seismic events,
- propagation of combustion products (generated either inside or outside the plant) through the ventilation system,
- possible water and steam ingress into vulnerable equipment,
- single random actuations of FPSs,
- multiple actuations of FPSs, and
- type of fire detectors.

4.1.2 Generic Plant Analysis

The safety significance of fire protection system (FPS) actuation with subsequent safety-related system damage is highly plant dependent (this is one reason why the IPEEE program was initiated). However, there are sufficient similarities among plants that some generic conclusions can be drawn. For this generic plant analysis, three types of fire zones were assessed: the cable spreading room(s), the emergency electrical switchgear room(s), and the diesel generator room(s). These zones were selected for several reasons. First, they are all represented in each of the individual plants studied (Refs. 4–8). Second, in each study, these zones are all contributors to overall risk. Third, these zones are representative of equivalent zones in all the U.S. commercial power plants. More directly, all power plants have spaces that are functional equivalents to cable spreading rooms, diesel generator rooms, and emergency electrical switchgear rooms. However, it must be emphasized that there may well be, in specific plants, other fire zones that dominate risk associated with fire suppressant damage to safety-related systems. Individual plant analysis must be done to identify such zones.

Fire suppressants studied include water, Halon and CO_2. Sensors used in the actuation control systems include:

- Smoke detectors, generally the ionization type. These sensors are sensitive to dust generated in a seismic event as well as fire-generated smoke.

- Flame detectors, used primarily in diesel generator rooms and in other areas where oil fires are likely.

- IR heat detectors.

- Cable tray heat detectors, that have long wire sensors in individual cable trays.

FPS piping systems are of two types; with pressurized and non-pressurized headers. In the generic cases presented, only the wet pipe water system is of the pressurized type. Water preaction, water deluge, Halon and CO_2 are the non-pressurized type, relying on sensors and the control system to detect the need for suppressant and subsequently actuate pressurization valves. Table 4.1 summarizes the elements of the various types of systems analyzed in the generic cases.

Different strategies have been employed in U.S. commercial nuclear power plants for critical fire zones as to the type of suppressant agent and fire protection system actuation scheme utilized. Table 4.2 lists the cases that were analyzed in this study.

Table 4.1 Fire Protection System Summary

System Type	Normally Pressurized	Sensors Utilized	Nozzle Type
H_2O Preaction	No	Yes	Fusible Link
H_2O Wetpipe	Yes	No	Fusible Link
H_2O Deluge	No	Yes	Open
Halon	No	Yes	Open
CO_2	No	Yes	Open

Table 4.2 Fire Protection Cases for Analysis

Suppression Agent	System Type	Cable Spreading Room	Diesel Generator Rooms	Emergency Switchgear Rooms
Water	Preaction	X	X	–
Water	Wetpipe	X	X	X
Water	Deluge	X	X	X
Halon		X	–	X
CO_2		X	X	X

X indicates analysis was performed

The conditions and assumptions used in the generic scenario analysis are discussed in detail in Reference 9. A data base of fire occurrences was developed in Reference 13 for a number of typical nuclear power plant fire zones. Those applicable to this generic issue are presented in Table 4.3.

Fragility data for seismic suppressant diversion are based on individual plant specific analysis. It must be noted that

while diversion is a concern in the assessment of Root Cause 12, it acts as an advantage in the analysis of Root Causes 7/8/9, because if suppressant is diverted away from a zone of interest, it cannot cause damage to safety-related components and cables.

Non-seismic fire protection system actuation frequencies (per system-year) are based on the LER data in Appendix A, and are provided in Table 4.4.

Table 4.3 Fire Occurrence Frequencies

Fire Zone	Frequency/Reactor-Year
Cable Spreading Room	2.7E–3
Diesel Generator Building	2.3E–2
Emergency Switchgear Room	3.0E–3

Thirteen generic cases were examined on the basis of these assumptions and conditions. They represent those cases presented in Table 4.2. Details of calculations are presented in Appendix E of Reference 9. Calculations for

core damage frequency, risk, and sensitivity studies were performed using the Top Event Matrix Analysis Code (TEMAC) (Ref. 14) and the Latin Hypercube Sampling Code (Ref. 15).

Table 4.4 FPS Actuation Frequency per System-Year

	Human Error	Hardware Failure	Other	Total
Water Preaction	5.2E–4	5.2E–4	5.2E–4	1.6E–3
Water Wet Pipe	2.8E–3	5.0E–3	1.8E–3	9.6E–3
Water Deluge	9.4E–3	5.5E–3	1.3E–2	2.8E–2
Halon	3.5E–3	5.3E–3	8.8E–4	9.7E–4
CO_2	1.4E–3	2.3E–3	1.8E–3	5.5E–3

4.1.3 Generic Core Damage Frequency Estimates

This section examines core damage frequency (CDF) in three generic plant cases. The following caveats must be kept in mind when reviewing the data presented in these cases:

- The generic plants examined are represented by only three fire zones: the cable spreading room, the diesel generator rooms, and the emergency electrical switchgear rooms. In any given specific plant, other fire zones may be significant or even dominant contributors to CDF, and specific plant analysis must be conducted to evaluate an individual plant (this was one of the reasons for initiation of the IPEEE program).

- Because the level of damage to electrical cables and electro-mechanical components that results from short term exposure to Halon FPS agent is not clear from experimental and historical data, Halon is not considered in evaluation of the best case plant. If the assumption is made that Halon presents no short term threat to cables and components, then the incremental CDF associated with a generic plant with all Halon FPS systems would be only that resulting from Root Cause 12 (seismic/fire interaction). Therefore, the CDF associated with Halon suppressant agent damage to cables and components would be zero.

- CDF data in the tables representing the generic cases are mean values for the individual root cause and room totals, and the sum of mean values for the overall plant values. Uncertainty calculations were not accomplished in examining the overall generic plant CDF values.

- In the typical and best generic plant cases, while in the CDF tables a value is presented for the emergency electrical switchgear rooms, it must be recognized that data (Appendix D to Reference 9) indicate that an automatic FPS is installed in these rooms in only about 20 percent of the U.S. commercial nuclear power plants. In the remaining 80 percent of these plants, there is no automatic FPS system installed. Thus for these generic cases, two overall values for CDF are provided, one for the case with an FPS in the emergency electrical switchgear rooms, and one for the case with no automatic FPS in those rooms. It is suspected that although Generic Issue 57 associated CDF for the case with no automatic FPS installed is lower, the overall plant CDF may not be lower because of the likelihood that the CDF associated with fire in these rooms is higher without an automatic FPS systems. A detailed analysis of this issue was not conducted.

- For all the cases, the LLNL seismic hazard curves were used.

4.1.3.1 Most Vulnerable Generic Plant

In the most vulnerable case plant, for each of the three rooms examined (cable spreading, diesel generator, and emergency electrical switchgear rooms) the FPS system resulting in the highest CDF was selected. In all cases, the resulting FPS system is deluge water. Additionally, for this case only, mercury wetted contact type relays were assumed to be installed in the deluge FPS control system. Data for this case are presented in Table 4.5. For the worst case generic plant, CDF is calculated to be 2.5E–3/year [of which 2.2E–3/year (88%) is due to the deluge water system in the diesel generator rooms]. This CDF is sufficiently high to require consideration of plant specific corrective actions on any plant whose configuration actually contains most or all of the assumed features of the hypothetical "most vulnerable" plant.

This hypothetical "most vulnerable" generic plant is assumed to be subject to the seismic hazard curve for the ANO site, including the recent New Madrid seismic event, and it is assumed that any seismic event from 1/2 SSE to 6 SSE will cause actuation of the plant's FPS (due to assumptions "a" and "b", described below for the FPS's configuration). This high CDF results from the FPS configuration assumptions that the plant either: a) has seismically sensitive relays in the FPS, or b) is actuated solely by a smoke sensitive detector; and c) that vulnerable components of the diesel generator are susceptible to water intrusion from the FPS.

Table 4.5 Core Damage Frequency – "Most Vulnerable" Generic Plant

Fire Zone: FPS Type: Root Cause	Cable Spreading Room Deluge Water	Diesel Generator Rooms Deluge Water	Emergency Electrical Switchgear Rooms Deluge Water
4	1.6E–6		1.8E–6
5	2.6E–8		3.1E–6
6	9.1E–7		1.1E–6
7	2.6E–5		
8	2.6E–5		
9			
7/8		2.2E–3	2.2E–4
8/9			
7/8/9			
11	4.4E–7		
12	1.2E–5		1.2E–5
13	2.3E–6		2.5E–6
Total	6.9E–5	2.2E–3	2.4E–4

Total CDF for "Most Vulnerable" Generic Plant: 2.5E–3/RY.

Efforts are underway to identify any actual plants with deluge water systems in the diesel generator rooms whose configuration meets assumptions (a) or (b) and (c); however, based on a preliminary survey, it is expected that very few (if any) will be identified in this category. Consideration of plant-specific followup will be given on a case-by-case basis.

4.1.3.2 Typical Case Generic Plant

In the typical case plant, for each of the three rooms examined, the FPS was chosen that represents the most common installation in commercial U.S nuclear power plants. Based on survey data (Appendix D to NUREG/CR–5580), there are 69 cable spreading rooms with wet-pipe water FPS systems, 49 diesel generator rooms with CO_2 FPS systems, and 18 switchgear rooms with CO_2 FPS systems. CDF data associated with these systems in the generic plant are shown in Table 4.6. For the typical case generic plant, CDF is calculated to be 3.5E–5/year with an automatic FPS in the emergency electrical switchgear rooms, and 2.4E–5/year with no automatic systems installed in the emergency electrical switchgear rooms.

4.1.3.3 Least Vulnerable Generic Plant

In the least vulnerable case plant several assumptions are made to optimize the plant for minimum CDF associated with Generic Issue 57. The assumptions are based on the information gained from the study. The assumptions are:

- For each of the three rooms examined, the FPS system resulting in the lowest CDF was selected. Accordingly, a CO_2 FPS was selected for the cable spreading room, a preaction water FPS was selected for the diesel generator rooms, and a CO_2 FPS was selected for the emergency electrical switchgear rooms.

- For the cable spreading room, it was assumed that there are no electrical cabinets in the room to act as fire sources in a seismic event. This assumption is consistent with some, but not all of the individual plants walked-down. This eliminates CDF associated with Root Cause 12 (seismic/fire interaction) in this space.

- For the emergency electrical switchgear rooms, the electrical cabinets that remain energized in a LOSP event are assumed to be seismically restrained against sliding or tipping, eliminating the CDF associated with Root Cause 12 (seismic/fire interaction). This kind of cabinet restraint was observed in some, but not all of the plants walked-down.

- For all three fire zones, relays in the FPS control systems are assumed to be seismically qualified, and the CDF associated with relay chatter is reduced by a factor of 10 from that in the typical generic plant case. Such seismically qualified relays were found in some, but not all, of the plants walked-down.

Data for this case are shown in Table 4.7. It should be noted that to achieve further reductions in CDF, the contributions due to non-seismic root causes must be reduced. The principal factors involved that must be reduced are the conditional probabilities for damage of cables and active electro-mechanical components, given that they are wetted by a fire suppressant agent. In this

Table 4.6 Core Damage Frequency – "Typical" Generic Plant

Fire Zone: FPS Type: Root Cause	Cable Spreading Room Wetpipe Water	Diesel Generator Rooms CO_2	Emergency Electrical Switchgear Rooms CO_2
4	4.7E–8		1.5E–8
5			2.6E–7
6	8.4E–8		2.5E–8
7			
8		1.2E–5	
9			
7/8			1.6E–6
8/9			
7/8/9			
11			
12	1.2E–5		9.4E–6
13	2.9E–8		2.1E–8
Total	1.2E–5	1.2E–5	1.1E–5

Total CDF for "Typical" Generic Plant: 3.5E-5/year with an automatic FPS installed in the emergency electrical switchgear room, 2.4E-5/RY without.

Table 4.7 Core Damage Frequency – "Least Vulnerable" Generic Plant

Fire Zone: FPS Type: Root Cause	Cable Spreading Room CO_2	Diesel Generator Rooms Preaction Water	Emergency Electrical Switchgear Rooms CO_2
4	2.3E–7		1.5E–8
5	2.6E–8		2.6E–7
6	3.9E–7		2.5E–8
7			
8			
9			
7/8	1.2E–7		1.6E–7
8/9			
7/8/9		< 1.0E–8	
11			
12	< 1.0E–8		< 1.0E–8
13	3.0E–7		2.1E–8
Total	1.1E–6	< 1.0E–8	4.8E–7

Total CDF for "Least Vulnerable" Generic Plant: 1.6E-6/RY with an automatic FPS installed in the emergency electrical switchgear room, 1.1E-6/RY without.

study, some of these values had to be established using zero data point bounding methods, while the remainder are based on very few documented actual damage events. A testing program could better define these conditional probabilities, and in all likelihood result in reduced calculated values for non-seismic root cause contributions to CDF.

For the "least vulnerable" case generic plant, CDF is calculated to be 1.6E–6/RY with an automatic FPS in the emergency electrical switchgear rooms, and 1.1E–6/RY with no automatic systems installed in the emergency electrical switchgear rooms.

Table 4.8 summarizes the CDF contributions from each root cause for the three plant-specific evaluations.

4.2 Dose Consequences Analysis

For purposes of this study, consequences are measured in person-sievert, abbreviated as person-Sv, with equivalent consequences in units of person-rem given enclosed in parentheses immediately following. Also, benefits are given in person-Sv (person-rem) averted. Once the CDF and changes in CDF from a potential resolution alternative have been calculated (Section 4.1), the next step is to calculate the corresponding consequences in person-Sv (person-rem), and hence, benefits in person-Sv (person-rem) averted.

Section 4.7 of Reference 9 contains a detailed discussion of the generic offsite dose calculations and presents generic risk values for each of the applicable root causes.

In addition to the plant-specific evaluations performed, three generic plant cases were analyzed representing design configurations and fire protection system type utilization in key plant locations so that "most vulnerable," "typical," and "least vulnerable" cases were identified. These assumed plant configurations may not necessarily correspond to the specific design of a given plant, but rather they represent a wide spectrum of possible designs, with the "typical" case judged to be the closest to the design and associated risk level of the predominant number of plants. The basis for this judgement is presented later in this section.

The generic plants examined are represented by only three fire zones: the cable spreading room, the diesel generator rooms, and the emergency electrical switchgear rooms. In any given specific plant, other fire zones may be significant or even dominant contributors to CDF and risk, and IPEEE plant specific analysis must be conducted to evaluate an individual plant.

Because the level of damage to electrical cables and electro-mechanical components that results from short term exposures to Halon FPS agent is not clear from experimental and historical data, Halon FPS systems are not considered in evaluation of the "least vulnerable case" plant. If the assumption is made that Halon presents no short term threat to cables and components, then the incremental CDF and risk associated with a generic plant with all Halon FPS systems would be only that resulting from Root Cause 12 (seismic/fire interaction). CDF and risk associated with Halon suppressant agent damage to cables and components would be zero.

Table 4.8 Plant-Specific Base Case Results in Terms of Core Damage Frequency
(Per Reactor Year)

Root Cause	Westinghouse Plant	B&W Plant	GE Plant
1.	N/A	N/A	5.7E–7
2.	N/A	N/A	N/A
3.	N/A	N/A	N/A
4.	1.4E–6	2.3E–6	3.3E–7
5.	1.1E–6	N/A	2.3E–8
6.	2.1E–6	1.4E–6	5.4E–7
7.	N/A	N/A	3.3E–7
8.	2.6E–7	1.5E–6	1.1E–5
9.	N/A	< 1.0E–8	N/A
10.	N/A	N/A	6.9E–7
11.	4.2E–7	6.4E–7	5.7E–7
12.	1.4E–6	4.7E–5	8.6E–6
13.	5.7E–7	2.9E–6	4.4E–7
Total	7.3E–6	5.6E–5	2.3E–5

CDF and risk data in the tables representing the generic cases are mean values for the individual root cause and room totals, and the sum of mean values for the overall plant values. Uncertainty calculations were not performed in examining the overall generic plant CDF and risk values. However, uncertainty could be expected to be distributed in a way similar to that in the specific plant analysis. From 5% to the 95% point in composite CDF, the range was about two orders of magnitude. For risk, the range would be expected to be considerably larger.

For purposes of this regulatory analysis the results of the evaluation of the "typical" plant case were used to assess the cost/benefit parameters associated with this issue.

In the typical case plant, for each of the three rooms examined, the FPS was chosen that represents the most common installation in commercial U.S. nuclear power plants. Based on survey data (Appendix D to Reference 9), there are 69 cable spreading rooms with wetpipe water FPS systems, 49 diesel generator rooms with CO_2 FPS systems, and 18 switchgear rooms with CO_2 FPS systems. Dose consequences data (PWR and BWR) associated with these systems in the generic plant are shown in Tables 4.9 and 4.10. For the typical case PWR, 20 year risk is 0.54 person-Sv (54 person-rem) with an automatic FPS installed in the emergency electrical switchgear room, and 0.51 person-Sv (51 person-rem) without. For the typical case BWR, 20 year risk 2.2 person-Sv (220 person-rem) with an automatic FPS installed in the emergency electri-

cal switchgear room, and 2.1 person-Sv (210 person-rem) without.

4.3 Cost Analysis

To calculate the cost for the various backfit alternatives the analysis drew from several sources and followed the guidelines of References 16 and 17. The computer code FORECAST 3.0 (Ref. 18) which incorporates the information in the preceding references was used to develop estimates for the various backfit alternatives. For each alternative the costs noted in Section 4.3.1 through 4.3.9 were considered.

4.3.1 Labor and Equipment/Materials Costs

The Energy Economic Data Base (EEDB), Reference 19 (embedded in the FORECAST code), and R.S. Means Cost Guides (Ref. 20) provided the basis for the equipment/material cost and labor estimates. The EEDB incorporates "as-built" cost information (both material unit cost and installation labor hours) for nuclear plant construction activities. The material and labor information from R.S. Means Cost required adjustment to the specified EEDB basis to properly reflect the nuclear plant level of effort and equipment/material specifications. Two factors, derived for and used in previous cost study (Ref. 21) were employed: Means-EEDB equipment/materials costs were adjusted by multiplying by 2.1 and Means-EEDB labor hours were adjusted by multiplying by 2.7. The cost modification factors of 2.1 for the equipment/material costs and 2.7 for the labor costs were utilized to

Table 4.9 Risk in person-Sv/RY (person-rem/RY) "Typical" PWR Plant

Fire Zone: FPS Type: Root Cause	Cable Spreading Room Wetpipe Water	Diesel Generator Rooms CO_2	Emergency Electrical Switchgear Rooms CO_2
4	3.6E–5 (3.6E–3)		1.1E–5 (1.1E–3)
5			2.0E–4 (2.0E–2)
6	6.4E–5 (6.4E–3)		1.9E–5 (1.9E–3)
7			
8		9.1E–3 (9.1E–1)	
9	4.0E–5 (4.0E–3)		
7/8			1.2E–3 (1.2E–1)
8/9			
7/8/9			
11			
12	9.1E–3 (9.1E–1)		7.2E–3 (7.2E–1)
13	2.2E–5 (2.2E–3)		1.6E–5 (1.6E–3)
Total	9.3E–3 (9.3E–1)	9.1E–3 (9.1E–1)	8.6E–3 (8.6E–1)

Total risk for "Typical" PWR Generic Plant: 0.027 person-Sv/RY (2.7 person-rem/RY) with an automatic FPS installed in the emergency electrical switchgear room, 0.026 person-Sv/RY (2.6 person-rem/RY) without.

Table 4.10 Risk in person-Sv/RY (person-rem/RY) "Typical" BWR Plant

Fire Zone: FPS Type: Root Cause	Cable Spreading Room Wetpipe Water	Diesel Generator Rooms CO_2	Emergency Electrical Switchgear Rooms CO_2
4	1.5E–4 (1.5E–2)		4.6E–5 (4.6E–3)
5			7.9E–4 (7.9E–2)
6	2.7E–4 (2.7E–2)		7.6E–5 (7.6E–3)
7			
8		3.7E–2 (3.7E+0)	
9	1.7E–4 (1.7E–2)		
7/8			4.9E–3 (4.9E–1)
8/9			
7/8/9			
11			
12	3.9E–2 (3.9E+0)		2.9E–2 (2.9E+0)
13	9.4E–5 (9.4E–3)		6.4E–5 (6.4E–3)
Total	3.9E–2 (3.9E+0)	3.7E–2 (3.7E+0)	3.5E–2 (3.5E+0)

Total risk for "Typical" BWR Generic Plant: 0.11 person-Sv/RY (11 person-rem/RY) with an automatic FPS installed in the emergency electrical switchgear room, 0.11 person-Sv/RY (11 person-rem/RY) without.

reflect costs incurred at nuclear power plants. The R.S. Means Cost Guide reflects costs for non-nuclear facilities and the costs must be modified to represent the level of effort and material/equipment specifications required at a nuclear power plant.

Additionally, for operating nuclear power plants there are a number of workplace characteristics which significantly reduce the level of productivity and thus increase the number of labor hours required to accomplish a task. These characteristics, discussed in detail in FORECAST 3.0, include access, congestion and interference, radiation, and task management. Since EEDB reflects only new (or "as-built") plant conditions, the installation labor hours were adjusted to properly consider actual conditions existing at operating nuclear plants.

The total labor costs associated with the proposed modifications include overhead charges (at 100 percent of direct labor) to account for contractor management, administrative support, rent, insurance, etc.

4.3.2 Engineering and Quality Assurance/Control Costs

These costs reflect the cost of engineering and design, as well as quality assurance/control (QA/QC) activities associated with implementing the requirements. For requirements affecting structures/systems already in-place (operating plants) the guidelines of abstract 6.4 of "Generic Cost Estimates," (Ref. 17) recommend that a 25 percent engineering and QA/QC factor be applied to the direct cost (i.e., labor and materials and cost but without any

overhead charges). All cost estimates developed in this study included this engineering and QA/QC cost component.

4.3.3 Radiation Exposure

Worker radiation exposure estimates were derived based on guidelines presented in Abstract 5.1 of Reference 17. The collective radiation exposure associated with the implementation of a proposed plant modification is estimated by taking the product of the in-field labor hours necessary to perform the task and the work area dose rate associated with that particular task.

In this study the work area in which the modifications would take place are considered to be either low-dose contaminated areas (cable vault/tunnel) or clean areas (diesel generator rooms). Therefore, radiation exposure is either minimal or zero for the modifications proposed in this study.

4.3.4 Health Physics Support Costs

Health physics requirements for the potential plant enhancements were developed based on information and guidelines presented in Abstract 2.1.6 of Reference 17. Two factors were considered; the size of the work crew and the magnitude of the radiation field. The plant health physics (HP) personnel perform radiation surveys that are conducted throughout the time required to perform the modification, staff radiological checkpoints, set up anticontamination clothing removal areas, as well as determine badging requirements.

Some of the modifications are performed in low radiation but contaminated work areas, such as the cable vault/tunnel. Therefore, the health physics support costs are highest for this type of improvement. However, a minimum health physics cost increment is associated even with physical modifications conducted in clean work areas since area radiological surveys and other HP activities still have to be performed.

4.3.5 Anti-Contamination Clothing Costs

Cost estimates for anti-contamination (anti-c) clothing used while performing the plant modifications were derived based on Abstract 2.1.5 of Reference 17. The cost per suitup assumes that each member of the work crew requires two complete sets of anti-c clothing per eight-hour shift. Included in the cost per suit-up area the cost of purchasing the anti-c clothing set, its wear-out rate, laundering costs, etc. Only work tasks conducted in contaminated plant areas were considered to include this cost increment.

4.3.6 Radioactive Waste Disposal Cost

The costs for disposal of radioactive wastes generated during plant modifications were derived based on guidelines of Abstract 2.1.4 of References 17. For the study the cost increment associated with the disposal of radioactive wastes is applicable only to those plant modifications that necessitated removal of existing system components located in a contaminated area. The costs are, however, insignificant (less than five percent of total cost).

4.3.7 Other Licensee Costs

Other costs incurred by the utility as a result of implementing the proposed physical plant modifications included the costs of re-writing procedures, training the staff (both maintenance and operations), and changing recordkeeping or reporting requirements. For each of the above stated cost categories, the costs were derived following the guidelines presented in Abstracts 2.2.2, 2.2.3, and 2.2.4, respectively, of Reference 17. In this study, for some of the plant modifications proposed, these costs represent a significant portion of the total cost.

4.3.8 NRC Costs

These cost represent NRC implementation costs. They account for such NRC activities as developing inspection guidelines and procedures, assuring compliance with the proposed regulatory action, and other technical tasks. In this study, the cost estimates associated with the NRC were primarily derived from guidelines and input provided by References 16 and 18.

4.3.9 Onsite Averted Costs

In addition to the costs associated with the modifications, an evaluation of the costs associated with the potential reduction of severe onsite consequences were evaluated. "A Handbook for Value-Impact Assessment" (Ref. 16) was used as the reference for performing this evaluation. The values for onsite averted cost were calculated using the following equation:

$$V_{op} = NU (F_o - F_n)$$

Where

V_{op} = the cost of avoided onsite property damage

N = the number of affected facilities

U = the present value of onsite property damage given a release

F_o = the original core damage frequency (base case)

F_n = the core damage frequency after implementing an option

and

$$U = \frac{c}{m} \cdot \left(\frac{e^{-rt_i}}{r^2}\right) \cdot (1 - e^{-r(t_f - t_i)}) \cdot (1 - e^{-rm})$$

where

c = cleanup, repair, and replacement power costs

t_f = years remaining until end of plant life

t_i = years before reactor begins operation

m = period of time over which damage costs are paid out

r = discount rate (for 10%, r = 0.10)

The cost handbook (Ref. 16) recommends best estimate values for input to calculating U (present value of onsite property damage given a release) as follows:

c = $1,650 x 10^6

m = 10 reactor-years

r = 0.10

t_f = 20 reactor-year

t_i = 0 reactor-years

Using the above values for calculating U yields the following result:

Best estimate – $9.0 billion/severe accident event

This value is then applied to the potential change in accident frequency, or these analyses, change in core damage frequencies for each option.

4.4 Cost Estimate Uncertainties

The areas of uncertainty associated with the cost estimating model for this study included the following:

1. Labor rate variations due to plant site location

2. Contingency allowance

3. Variability of in-plant work environment conditions

4. Licensee procedural/administrative/analytical conditions

5. NRC procedural/administrative/analytical cost

6. Discount rate variation in the recurring cost module

7. Waste disposal cost module

Each cost estimate was evaluated to determine all areas of uncertainty applicable. Specific numerical values were used for each individual cost analysis. More detailed discussions of cost uncertainties may be found in Section 5.1.6 of Reference 9.

4.5 Backfit Alternatives Cost Estimates

A discussion of the proposed backfit alternatives based on plant walkdowns and evaluations is presented in the following subsections.

4.5.1 Alternative 1 – No Action

Under this alternative resolution there will be no cost involved.

4.5.2 Alternative 2 – Upgrade a FPS Actuation Controller with Seismically Qualified Printed Circuit Boards

Because of concerns and industry experience with relay chatter during seismic events it may be prudent to investigate replacing existing relays in the FPS controller cabinets with printed circuit boards. Based on plant specific walkdowns many types of FPS actuation relays were found. For some plants, mercury wetted relays for FPS actuation and/or the annunciation of alarms and isolation of room cooling was found. Given a seismic event, there is a high likelihood of actuation for some of these relay types. The intent of this modification is to replace the relays with printed circuit boards and prevent any damage which may result from inadvertent FPS actuation during a seismic event. For each of the areas modified, the contribution to the core damage frequency from Root cause 8 (relay chatter in a seismic event) would be eliminated.

For the purpose of this estimate, it is assumed that this modification could be performed during a planned unit outage. Therefore, costs associated with unit shutdown or startup and replacement power costs are not included in the estimate. It is assumed that this type of activity has been done many times before, requiring no learning curve adjustments. No significant radwaste disposal is involved. Also, the costs for security and fire watch personnel are estimated.

The total cost to implement this plant modification ranges from $13,000 to $17,000.

4.5.3 Alternative 3 – Replace Smoke Detector Actuated FPS with Heat Detector Actuated System

Replacing an existing smoke detector actuated system with a heat detector actuated system will eliminate contributions from Root Cause 1, 7 and 10 which are smoke detector specific. However, to provide an additional detection capability, it may be prudent to leave the existing smoke detectors intact for indication purposes only.

As with the upgrade of FPS controllers in the previous section, it is assumed that this project could be completed during a scheduled outage. No costs associated with shutdown, startup, and replacement power are included. A general set of productivity factors representative of a cable spreading room is used. Cost for security personnel and a fire watch (306 person-hours) are included in the total installation labor cost.

The cost to implement this option would range from $78,000 to $105,000.

4.5.4 Alternative 4 – Reroute Safety-Related Cables

Through plant walkdowns it had been found that there are certain "pinch points" located in the plant where cabling for certain redundant safety-related systems are run together. Given a fire or FPS actuation that could damage these cables, these safety systems are vulnerable to simultaneous failure. Therefore the intent of this modification is to reroute one of the sets of cabling to remove it from a common failure vulnerability.

For the purposes of this estimate, it is assumed that the old cable run would be eliminated in place and that the new cable installation could be completed during a planned unit outage. Therefore costs associated with unit shutdown or startup and replacement power costs are not included in the estimate. It is further assumed that one length of cable would be required (only control) and that the cable would need to be qualified for harsh environments. Thus, an estimate is made for cable subject to the requirements of the plant equipment qualification program. Costs for cable in conduit are used for all of the cable runs. The total length of cable run in conduit required is assumed to be approximately 152 meters (500 feet). Ten penetrations are required, and terminations are needed at both ends of the cable run.

Separate environmental and labor productivity factors are used for the two main plant areas. It is assumed that this type of activity has been done many times before, requiring no learning curve adjustments. It is assumed that part of the rerouting will be done in a radiation area and as such appropriate factors and costs for anti-c clothing and HP support are included. No significant radwaste disposal is involved. Also, the costs for security and fire watch personnel are estimated.

The total cost to implement this plant modification ranges from $136,000 to $185,000.

4.5.5 Alternative 5 – Seismically Qualify the CO_2 Tank, Outlet Piping, and Battery Rack

A seismic walkdown and subsequent fragility analysis for a typical CO_2 system found a high likelihood of suppressant agent diversion given an earthquake. Failure of the tank or its outlet piping or the FPS battery dominated the overall probability of failure of the CO_2 system. This potential plant modification would seismically qualify the CO_2 tank, battery, and its immediate outlet piping.

For the modification, it is assumed that this project would not require a special plant shutdown to implement. Therefore, shutdown, startup and replacement power costs are not included. Given that there is no real potential for contamination, cost estimates for health physics support and anti-c clothing are not included. Labor hours were included for security and fire watch support.

The total cost for this upgrade ranges from $97,000 to $131,000.

4.5.6 Alternative 6 – Seismically Qualify A FPS Battery Rack

It was found during a plant specific walkdown, that a water based FPS for an entire plant had two electric driven fire pumps and one diesel driven fire pump. Given a LOSP, both electric driven pumps would fail due to being powered from non-vital busses. A failure of the starting battery for the diesel pump would lead to a loss of a fire main pressure.

For this modification it is assumed that this project would not require a special plant shutdown to implement. Therefore, shutdown, startup and replacement power costs are not included. Given that there is not a real potential for contamination, cost estimates for health

physics support and anti-c clothing are not included. Labor hours are included for security and fire watch support.

The total cost for this ranges from $35,000 to $42,000.

4.5.7 Alternative 7 – Upgrade the FPS Water Quality

Water is the most frequently used fire suppressant at nuclear power plants. The most common delivery systems are pre-action sprinklers, deluge wet-pipe sprinklers, dry-pipe sprinklers and standpipe hose systems. It has been postulated that improvement of water quality will reduce the potential for damage to safety-related components from exposure to water suppressant. This is based on the thought that pure water, less conductive than normal fire fighting water, would be less likely to cause short circuits or grounds. To address this concern, a modification to upgrade the FPS water quality is performed. For this modification a water purification system would be required along with the associated piping and valves and a storage tank. It is assumed for this modification that the existing FPS piping and pumps would be utilized. Also, it is assumed that this project could be performed during plant operation. Therefore, shutdown, startup and replacement power costs are not included. Given that there is not a real potential for contamination, costs estimates for health physics support and anti-c clothing are not included. Labor are hours included for security and fire watch support.

The total cost for this upgrade ranges from $1,174,000 to $1,577,000.

4.5.8 Alternative 8 – Replace Deluge with Preaction Sprinkler FPS

Of all the types of water based FPS utilized in nuclear power plants the pre-action sprinkler system has been found to experience the least number of inadvertent actuations (Ref. 9). A pre-action sprinkler system requires the opening of a deluge valve, either automatically, with a control signal, or manually, and a rise in the ambient temperature to the melting point of the fusible links on the sealed sprinkler heads. A deluge FPS only requires the opening of a valve, either automatically or manually to discharge water under pressure to the open spray heads. This modification involves replacing a deluge FPS with a pre-action sprinkler FPS. This modification would reduce the frequency of inadvertent actuations and localize the application of FPS agent.

For this modification fusible link sealed sprinkler heads would need to be added to the existing deluge FPS. All of the existing hardware would be kept in place. It is also assumed the existing locations of sprinkler heads would be adequate for preaction system. This modification will not require a special plant shutdown to implement. Therefore, shutdown, startup and replacement power costs are not included. Also this modification will not require any health physics support or anti-c clothing. Labor hours are included for security and fire watch support.

The total cost for this upgrade ranges from $22,000 to $30,000.

4.5.9 Alternative 9 – Replace Electrical Cabinet with a Cabinet Designed to Prevent Water Intrusion

One study (Ref. 11) has reported that electrical equipment failure modes related to water were mainly due to electrical shorting and long term corrosion. Wherever water intrusion is possible the equipment could be expected to fail through shorting, grounding, tripping of overcurrent devices, physical damage due to the velocity of direct hose streams, or long term corrosion causing potential failure of electro-mechanical parts. These failure modes were found to be dependent on the national electrical manufacturers associated (NEMA) rating or the configuration of the electrical enclosures. The appropriate NEMA rating to preclude water intrusion can potentially eliminate failures in electrical equipment due to water based FPSs. Enclosures that have a NEMA rating of 1 and 5 are subject to water intrusion under all water spray conditions applicable for this study. Enclosure with NEMA ratings of 2, 3, 3R, 3S, 4, 4S, 6, 6P, 11, 12, 12K and 13 are expected to prevent water intrusion under direct hose stream and splashing water spray. Only those enclosures with a NEMA rating of 6 and 6P are expected to prevent water intrusion under temporary submersion due to flooding. The intent of this modification is to replace existing safety-related electrical cabinets and enclosures with NEMA spray-proof rated enclosures. Although the modified electrical cabinets may require internal cooling

a cost for internal cooling was not included for this analysis.

Unlike the other generic modifications presented in this section this modification can not be completed during a scheduled outage since safety-related cabinets would need to be de-energized. Therefore, costs associated with shutdown, startup and replacement power are included. Additionally, costs for health physics support and anti-c clothing are included. Costs for security and fire watch are also included. The cost to implement this modification would range from $22,000 to $30,000 per cabinet.

4.5.10 Alternative 10 – Seismically Anchor Safety-Related Cabinets Susceptible to Tipping/Sliding Failure

In a seismic event energized cabinets may present a potential source for fire. Although it is assumed that in a seismic event offsite power will be lost, thus deenergizing many electrical cabinets, there will be a number of safety-related cabinets energized by alternative power sources (batteries, diesel generators). These energized cabinets are susceptible to tipping/sliding failure possible leading to a fire.

It is assumed that this modification would eliminate the potential for seismically induced fires due to the tipping/sliding failure of an energized electrical cabinet. For this modification a plant shutdown will not be required. Therefore, shutdown, startup and replacement power costs are not included. Additionally, it is assumed there is no real potential contamination so that cost estimates for health physics support and anti-c clothing are not included. Labor hours are included for security watch support. The total cost for this modification ranges from $67,000 to $91,000.

5 VALUE/IMPACT ANALYSIS

The value/impact (V/I) methodology for analyzing the various alternatives examined under this study is based on the requirements of the backfit rule (10 CFR 50.109) and related implementing guidance contained in References 16, 22, and 23. One of the primary considerations here is the derivation of cost/benefit ratios for each alternative evaluated in terms of cost in dollars per person-Sv (dollars per person-rem) averted, which may be compared to a guideline such as $100,000 per person-Sv ($1,000 per person-rem). This quantitative guidance is one of the elements considered in the decision-making process. Also, the interim guidance contained in Reference 10 is consistent with the objectives set forth in Section 2. Deterministic considerations on the merits of a proposed alternative resolution are also part of the decision with respect to a given alternative (Section 6). In the following subsections a description of each alternative and the results of a value/ impact assessment are presented for backfit as well as frontfit cases.

5.1 Backfit Analysis

Potential backfits, were identified as a result of plant walkdowns and evaluations and those analyzed are presented here as alternatives to resolve this issue.

5.1.1 Alternative 1 – No Action

Under this alternative there would be no new regulatory requirements. Consistent with existing regulations, this alternative does not preclude a licensee, or an applicant for a license, from proposing to the NRC staff design changes intended to enhance the reliability or operability of the fire protection systems and their components on a plant-specific basis.

5.1.2 Alternative 2 – Upgrade an FPS Actuation Controller with Seismically Qualified Printed Circuit Boards

A cost for this modification of $14K (one FPS controller) was estimated. This modification is assumed to eliminate any contribution of risk from Root Cause 8 or any combination of Root Cause 8 scenarios. It is assumed that for the cable spreading room (CSR) one FPS controller will be replaced and for the emergency diesel generator (EDG) and emergency switchgear (ESGR) areas two FPS controllers will be replaced.

Table 5.1 presents the cost/benefit results for this modification. Even without OSAC included this option appears to be cost-beneficial for EDG areas with a deluge or CO_2 system and for the ESGR with a deluge FPS. When OSAC are included a Halon FPS in the CSR for both a PWR and BWR and a deluge and CO_2 FPS in the CSR for BWR appears to be cost-beneficial.

5.1.3 Alternative 3 – Replace Smoke Detector Actuated FPS with a Heat Detector Actuated System

A cost for this modification of $87K was estimated. This modification was proposed only for the cable spreading room and would eliminate any incremental contribution from Root Cause 7 to core damage frequency from this area. The risk reduction for this modification is 0.45 person-Sv (45 person-rem) for the PWR analyzed and 1.0 person-Sv (180 person-rem) for the BWR analyzed. Table 5.2 presents the cost/benefit results for this modification. If OSAC are not included, this modification appears to be beneficial only for the BWR analyzed. If OSAC are included both the PWR and BWR analyzed indicate a beneficial option.

5.1.4 Alternative 4 – Reroute Safety-Related Cables

A cost for this modification of $154K was estimated. This modification is assumed to apply only to the cable spreading room and was assumed to reduce risk in the CSR by an order of magnitude. If OSAC are not included this modification appears to only be beneficial for a deluge FPS in a BWR. With OSAC included, a deluge FPS in both a PWR and a BWR and a wetpipe and preaction FPS in a BWR indicate a beneficial modification. Table 5.3 presents the cost/benefit results for this modification.

5.1.5 Alternative 5 – Seismically Qualify the CO_2 Tank, Outlet Piping, and Battery Rack

For this modification it was assumed that there is one common CO_2 tank per plant. A cost for this modification of $109K was estimated. Implementing this modification eliminates the incremental contribution to core damage frequency from Root Cause 12 (seismic/fire interaction) for CO_2 systems. When OSAC are included for the cable spreading room and emergency switchgear room for the BWR analyzed, a beneficial modification is indicated. Table 5.4 presents the cost/benefit results for this modification.

Table 5.1
Alternative 2
$K/person·Sv ($K/person·rem) averted

	Preaction	Deluge	Wetpipe	CO_2	Halon
PWR					
CSR	6.7E+04 (6.7E+02)	7.8E+02 (7.8E+00)		7.0E+02 (7.0E+00)	6.4E+02 (6.4E+00)
EDG	2.5E+05 (2.5E+03)	1.9E+01 (1.9E–01)		9.0E+01 (9.0E–01)	2.2E+03 (2.2E+01)
ESGR		1.0E+02 (1.0E+00)		1.1E+03 (1.1E+01)	2.2E+03 (2.2E+01)
CSR	**6.6E+04** (**6.6E+02**)	**2.3E+02** (**2.3E+00**)		**1.6E+02** (**1.6E+00**)	**1.0E+02** (**1.0E+00**)
EDG	**2.5E+05** (**2.5E+03**)	**<1.0E+02** (**<1.0E+00**)		**<1.0E+02** (**<1.0E+00**)	
ESGR		**<1.0E+02** (**<1.0E+00**)		**5.2E+02** (**5.2E+00**)	**1.6E+03** (**1.6E+01**)
BWR					
CSR	1.6E+04 (1.6E+02)	1.9E+02 (1.9E+00)		1.8E+02 1 (1.8E+.00)	1.6E+02 (1.6E+00)
EDG	6.7E+04 (6.7E+02)	4.9E+00 (4.9E–02)		2.3E+01 (2.3E–01)	
ESGR	2.5E+01 (2.5E–01)	2.5E+02 (2.5E+00)		5.6E+02 (5.6E+00)	
CSR	**1.6E+04** (**1.6E+02**)	**5.5E+01** (**5.5E–01**)		**4.0E+01** (**4.0E–01**)	**2.6E+01** (**2.6E–01**)
EDG	**6.7E+04** (**6.7E+02**)	**<1.0E+02** (**<1.0E+00**)		**<1.0E+02** (**<1.0E+00**)	
ESGR		**<1.0E+02** (**<1.0E+00**)		**1.2E+02** (**1.2E+00**)	**4.2E+02** (**4.2E+00**)

Bold type indicates OSAC included

5.1.6 Alternative 6 – Seismically Qualify A FPS Battery Rack

The cost for this modification was estimated to be $39K. This modification is assumed to eliminate Root Cause 12 core damage frequency contributions from water-based FPSs. Table 5.5 presents the cost/benefit results for this modification. For the BWR examined a beneficial modification is indicated for preaction, deluge and wetpipe FPSs in the cable spreading room. For the emergency switchgear room a beneficial modification is indicated for deluge and wetpipe FPS. If OSAC are included all water-based FPSs considered (both PWR and BWR) would indicate a cost-beneficial modification.

5.1.7 Alternative 7 – Upgrade the FPS Water Quality

A cost for this modification of $1314K was estimated. The magnitude of risk reduction that would be achieved by improvement of water quality is not clear. Therefore a cost/benefit ratio was not calculated. Further study on the effect of water-based FPSs on safety-related equipment and the improvement of the FPS water quality and its effect may provide the data necessary to examine this issue quantitatively.

Table 5.2 Cost/Benefit Ratio
Alternative 3
$K/person·SV ($K/person·rem) averted

	Preaction	Deluge	Wetpipe	CO_2	Halon
PWR					
CSR		1.9 E + 02 (1.9E + 00)			
EDG					
ESGR					
CSR		< 1.0E + 02 (< 1.0E + 00)			
EDG					
ESGR					
BWR					
CSR		4.8E + 01 (4.8E–01)			
EDG					
ESGR					
CSR		< 1.0E + 02 (< 1.0E + 00)			
EDG					
ESGR					

Bold type indicates OSAC included.

Table 5.3 Cost/Benefit Ratio
Alternative 4
$K/person·Sv ($K/person·rem) averted

	Preaction	Deluge	Wetpipe	CO_2	Halon
PWR					
CSR	1.1E + 03) (1.1E + 01	3.0E + 02 (3.0E + 00)	1.1E + 03 (1.1E + 01)	1.5E + 03 (1.5E + 01)	4.0E + 03 (4.0E + 01)
EDG					
ESGR					
CSR	**3.2E + 02** **(3.2E + 00)**	**< 1.0E + 02** **(< 1.0E + 00)**	**3.2E + 02** **(3.2E + 00)**	**4.8E + 02)** **(4.8E + 00**	**2.1E + 03** **(2.1E + 01)**
EDG					
ESGR					
BWR					
CSR	2.8E + 02 (2.8E + 00)	7.4E + 01 (7.4E–01)	1.9E + 02 (1.9E + 00)	2.9E + 02 (2.9E + 00)	5.4E + 02 (5.4E + 00)
EDG					
ESGR					
CSR	**7.9E + 01** **(7.9E–01)**	**< 1.0E + 02** **(< 1.0E + 00)**	**5.3E + 01** **(5.3E–01)**	**9.5E + 01** **(9.5E–01)**	**2.8E + 02** **(2.8E + 00)**
EDG					
ESGR					

Bold type indicates OSAC included

Table 5.4 Cost/Benefit Ratio
Alternative 5
$K/person·Sv ($K/person·rem) averted

	Preaction	Deluge	Wetpipe	CO$_2$	Halon
PWR					
CSR				7.3E+02 (7.3E+00)	
EDG					
ESGR				6.8E+02 (6.8E+00)	
CSR				**1.8E+02 (1.8E+00)**	
EDG					
ESGR				**1.5E+02 (1.5E+00)**	
BWR					
CSR				1.8E+02 (1.8E+00)	
EDG					
ESGR				1.8E+02 (1.8E+00)	
CSR				**4.4E+01 (4.4E-01)**	
EDG					
ESGR				**4.0E+01 (4.0E-01)**	

Bold type indicates OSAC included

5.1.8 Alternative 8 – Replace Deluge with Preaction Sprinkler FPS

This modification applies to all three plant areas examined and would require one FPS replacement in the CSR and two in both the EDG and the ESGR area. However, a cost/benefit was not performed for the ESGR since a preaction FPS was not part of the configurations studied based on Appendix D of Reference 9. The cost for this modification was estimated to be $25K for one plant area. The reduction in risk associated with this modification is the difference in risk between a deluge FPS and a preaction FPS. A cost-beneficial modification is indicated for both the cable spreading room and the emergency diesel generator area without OSAC included for both the PWR and BWR examined. Table 5.6 presents the cost/benefit results for this modification.

5.1.9 Alternative 9 – Replace Electrical Cabinet with a Cabinet Designed to Prevent Water Intrusion

This modification was examined on a plant area basis assuming two cabinets in each of two EDG areas, two cabinets in the CSR and ten cabinets in each of two ESGR areas. These numbers were estimated based on plant walkdowns and may vary for specific plants. The cost for this modification was estimated to be $25K for one cabinet replacement. The reduction in risk associated with this modification is the elimination of all water-based core damage frequency contributions except for Root Cause 12. For the deluge FPS this modification appears to be cost-beneficial in the cable spreading room and emergency diesel generator areas for both the PWR and BWR examined without OSAC included. If OSAC are included, this modification also appears to be beneficial for a wet-pipe FPS in an EDG area in a BWR. The results of the cost/benefit are presented in Table 5.7.

	Preaction	Deluge	Wetpipe	CO$_2$	Halon
PWR					
CSR	2.0E + 02 (2.0E + 00)	2.0E + 02 (2.0E + 00)	2.0E + 02 (2.0 E + 00)		
EDG					
ESGR		< 1.0E + 02 (< 1.0E + 00)	< 1.0E + 02 (< 1.0E + 00)		
CSR					
EDG					
ESGR					
BWR					
CSR	4.8E + 01 (4.8E–01)	4.8E + 01 (4.8E–01)	4.8E + 01 (4.8 E–01)		
EDG					
ESGR		< 1.0E + 02 (< 1.0E + 00)	< 1.0E + 02 (< 1.0 E + 00)		
CSR					
EDG					
ESGR					

Bold type indicates OSAC included

	Preaction	Deluge	Wetpipe	CO$_2$	Halon
PWR					
CSR		4.5E + 01 (4.5E–01)			
EDG		3.3E + 01 (3.3E–01)			
ESGR					
CSR		< 1.0E + 02 (< 1.0E + 00)			
EDG		< 1.0E + 02 (< 1.0E + 00)			
ESGR					
BWR					
CSR		1.1E + 01 (1.1E–01)			
EDG		8.8E + 00 (8.8E–02)			
ESGR					
CSR		< 1.0E + 02 (< 1.0E + 00)			
EDG		< 1.0E + 02 (< 1.0E + 00)			
ESGR					

Bold type indicates OSAC included.

Table 5.7 Cost/Benefit Ratio
Alternative 9
$K/person-Sv ($K/person-rem) averted

	Preaction	Deluge	Wetpipe	CO$_2$	Halon
PWR					
CSR	8.1E+04 (8.1E+02)	9.1E+01 (9.1E–01)	1.6E+04 (1.6E+02)		
EDG	9.1E+05) (9.1E+03	6.7E+01 (6.7E–01)	3.4E+05 (3.4E+03)		
ESGR		1.3E+03 (1.3E+01)	1.7E+04 (1.7E+02)		
CSR	**8.0E+04 (8.0E+02)**	**<1.0E+02 (<1.0E+00)**	**1.5E+04 (1.5E+02)**		
EDG	**9.1E+05 (9.1E+03)**	**<1.0E+02 (<1.0E+00)**	**3.4E+05 (3.4E+03)**		
ESGR		**6.8E+02 (6.8E+00)**	**1.6E+04 (1.6E+02)**		
BWR					
CSR	2.0E+04 (2.0E+02)	2.3E+01 (2.3E–01)	1.3E+02 (1.3E+00)		
EDG	2.4E+05 (2.4E+03)	1.8E+01 (1.8E–01)	1.0E+05 (1.0E+03)		
ESGR		3.2E+02 (3.2E+00)	4.2E+03 (4.2E+01)		
CSR	**2.0E+04 (2.0E+02)**	**<1.0E+02 (<1.0E+00)**	**1.3E+02 (1.3E+00)**		
EDG	**2.4E+05 (2.4E+03)**	**<1.0E+02 (<1.0E+00)**	**1.8E+01 (1.8E–01)**		
ESGR		**1.7E+02 (1.7E+00)**	**4.0E+03 (4.0E+01)**		

Bold type indicates OSAC included.

5.1.10 Alternative 10 – Seismically Anchor Safety-Related Cabinets Susceptible to Tipping/Sliding Failure

This modification applies to the CSR and ESGR areas and the cost appropriately reflects the number of electrical cabinets in each area. It is recognized that some of the cabinets may already be seismically anchored, but for the purposes of this estimate it is assumed that all cabinets will require the seismic anchoring modification. The cost for seismically anchoring one cabinet is estimated to be $76K. The reduction in risk associated with this modification is the elimination of Root Cause 12 scenarios in the CSR and ESGR areas. Table 5.8 presents the cost/benefit results for this modification. This modification appears to be beneficial when OSAC are included for a preaction, deluge and wetpipe FPS in the CSR for the BWR examined.

5.2 Frontfit Analysis

The plant modifications presented in the cost/benefit analysis as part of the Generic Issue 57 generic plant analysis were intended to be backfits for existing plants and as such these proposed plant modifications were determined from the insights provided by the three individual plant analyses and the generic plant analysis. However, some of these plant modifications may be considered as frontfits as part of the Advanced Light Water Reactor (ALWR) design program. All of the modifications considered as frontfits would avoid cost contributions for health physics support, radiation exposure, waste disposal and removal labor. Licensee and NRC costs were not recalculated and are not expected to differ significantly from the backfit estimates. The following subsections specifically discuss the proposed plant modifications, as applicable, for frontfits.

Table 5.8 Cost/Benefit Ratio
Alternative 10
$K/person·Sv ($K/person·rem) averted

	Preaction	Deluge	Wetpipe	CO$_2$	Halon
PWR					
CSR	7.6E+02 (7.6E+00)	7.6E+02 (7.6E+00)	7.6E+02 (7.6E+00)	1.0E+03 (1.0E+01)	2.8E+03 (2.8E+01)
EDG					
ESGR		7.6E+03) (7.6E+01	7.6E+03 (7.6E+01)	9.5E+03 (9.5E+01)	2.2E+04 (2.2E+02)
CSR	**2.2E+02 (2.2E+00)**	**2.2E+02 (2.2E+00)**	**2.2E+02 (2.2E+00)**	**4.7E+02 (4.7E+00)**	**2.3E+03 (2.3E+01)**
EDG					
ESGR		7.1E+03 (7.1E+01)	7.1E+03 (7.1E+01)	9.0E+03 (9.0E+01)	2.1E+04 (2.1E+02)
BWR					
CSR	1.9E+02 (1.9E+00)	1.9E+02 (1.9E+00)	1.9E+02 (1.9E+00)	2.5E+02 (2.5E+00)	6.9E+02 (6.9E+00)
EDG					
ESGR		2.1E+03 (2.1E+01)	2.1E+03 (2.1 E+01)	2.5E+03 (2.5E+01)	6.1E+03 (6.1E+01)
CSR	**5.4E+01 (5.4E–01)**	**5.4E+01 (5.4E–01)**	**5.4E+01 (5.4 E–01)**	**1.1E+02 (1.1E+00)**	**5.6E+02 (5.6E+00)**
EDG					
ESGR		1.9E+03 (1.9E+01)	1.9E+03 (1.9 E+01)	2.4E+03 (2.4E+01)	5.9E+03 (5.9E+01)

Bold type indicates OSAC included.

5.2.1 Alternative 1 – No Action

Under this alternative there would be no new regulatory requirements. Consistent with existing regulations, this alternative does not preclude an applicant under 10 CFR Part 50 or Part 52 from proposing to the NRC staff design changes intended to enhance the reliability or operability of the fire protection systems and their components on a plant design-specific basis.

5.2.2 Alternative 2 – Upgrade a FPS Actuation Controller with Seismically Qualified Printed Circuit Boards

Because of concerns and industry experience with relay chatter during seismic events it may be prudent to investigate replacing existing relays in the FPS controller cabinets with printed circuit boards. Based on plant specific walkdowns many types of FPS actuation relays were found. For some plants, mercury wetted relays for FPS actuation and/or the annunciation of alarms and isolation of room cooling was found. Given a seismic event, there is

a high likelihood of actuation for some of these relay types. The cost determined for this design is for one FPS actuation controller and would eliminate Root Cause 8 (relay chatter in a seismic event) contributions in the area of installation only. If this modification is performed as a frontfit it would be included as part of the overall FPS design. However, it is assumed that the costs associated with including this relay type as part of a new design would be similar to that of the backfit costs minus the costs of the original relay. The total cost of this relay ranges from $13,000 to $17,000.

5.2.3 Alternative 3 – Replace Smoke Detector Actuated FPS with a Heat Detector Actuated FPS

Designing a FPS to actuate on heat detectors rather than smoke detectors will eliminate contributions from Root Cause 1, 7 and 10 which are smoke detector specific. The cost for this system as a frontfit would be similar to the costs estimated for the components considered as part of an existing plant backfit. The cost to implement this option would range from $78,000 to $105,000.

5.2.4 Alternative 4 – Reroute Safety-Related Cables

The intent of this modification is to reroute one set of redundant cabling to remove it from a common mode failure vulnerability. This plant modification would not be considered as a frontfit and would be proposed as part of the new plant design.

5.2.5 Alternative 5 – Seismically Qualify the CO_2 Tank, Outlet Piping, and Battery Rack

A seismic walkdown and subsequent fragility analysis for a typical CO_2 system found a high likelihood of suppressant agent diversion given an earthquake. Failure to the tank or its outlet piping or the FPS battery dominated the overall probability of failure of the CO_2 system. This potential plant system design would seismically qualify the CO_2 tank, battery, and its immediate outlet piping. The overall costs for this frontfit would not differ from the backfit costs significantly. The total cost for this design ranges from $97,000 to $131,000.

5.2.6 Alternative 6 – Seismically Qualify a FPS Battery Rack

The cost for this system as a frontfit would be similar to the costs estimated for an existing plant backfit. The total cost for this upgrade ranges from $35,000 to $47,000.

5.2.7 Alternative 7 – Upgrade the FPS Water Quality

Water is the most frequently used FPS agent at nuclear power plants. The most common delivery systems are pre-action sprinklers, deluge wet-pipe sprinklers, dry-pipe sprinklers and standpipe hose systems. It has been postulated that improvement of water quality will reduce the potential for damage to safety-related components from exposure to water suppressant. To address this concern, a plant design to upgrade the FPS water quality is considered. For this design a water purification system would be required along with the associated piping and valves and a storage tank. Additional costs to be considered for a frontfit would be the FPS sprinkler piping and sprinkler heads. The total cost for this system design would range from $1,174,000 to $1,577,000.

5.2.8 Alternative 8 – Replace Deluge with Preaction Sprinkler FPS

This modification would not be considered a frontfit, but part of an overall FPS design. However, it is anticipated that for the components of the FPS analyzed as part of the backfit analysis utilized as part of the new FPS design the costs will be similar.

5.2.9 Alternative 9 – Replace Electrical Cabinet with a Cabinet Designed to Prevent Water Intrusion

One study (Ref. 11) has reported that electrical equipment failure modes related to water were mainly due to electrical shorting and long term corrosion. Wherever water intrusion is possible the equipment could be expected to fail through shorting, grounding, tipping of overcurrent devices, physical damage due to the velocity of direct hose steams or long term corrosion causing potential failure of electro-mechanical parts. These failure modes were found to be dependent on the National Electrical Manufactures Associated (NEMA) rating or the configuration of the electrical enclosure. The appropriate NEMA rating to preclude water intrusion can potentially eliminate failures in electrical equipment due to water based FPSs. Enclosures that have NEMA rating of 1 and 5 are subject to water intrusion under all water spray conditions applicable for this study. Enclosures with NEMA ratings of 2, 3, 3R, 3S, 4, 4S, 6, 6P, 11, 12, 12K and 13 are expected to prevent water intrusion under direct hose stream and splashing water spray. Only those enclosures with a NEMA rating of 6 and 6P are expected to prevent water intrusion under temporary submersion due to flooding. The intent of this modification is to replace existing safety-related electrical cabinets and enclosures with NEMA spray-proof rated enclosures.

The cost for this system as a frontfit would be similar to the costs estimated for an existing plant backfit with the exception of the labor for the removal of old electrical cabinets. The cost to implement this modification as a frontfit would range from $22,000 to $30,000 per cabinet.

5.2.10 Alternative 10 – Seismically Anchor Safety-Related Cabinets Susceptible to Tipping/Sliding Failure

In a seismic event energized cabinets present a potential source for fire. Although it is assumed that in a seismic event offsite power will be list, thus deenergizing many electrical cabinets, there will be a number of safety-related cabinets energized by alternative power sources (batteries, diesel generators). These energized cabinets are susceptible to tipping/ sliding failure possibly leading to a fire. It is assumed that this modification would eliminate the potential for seismically induced fires due to the tipping/sliding failure of an energized electrical cabinet. The cost for this system as a frontfit would be similar to the costs estimated for an existing plant backfit. The total cost for this design ranges from $67,000 to $91,000.

Additional technical insights for the ALWR and the potential risks associated with the actuation of fire protection systems are given in some detail in Chapter 6 of Reference 9.

6 DECISION RATIONALE

Generic Issue 57, "Effects of Fire Protection System Actuation on Safety-Related Equipment", was identified in 1982 (Refs. 1,2) as a result of a number of precursor events showing that safety-related equipment subjected to fire protection system (FPS) water spray could be rendered inoperable. These precursor events also indicated numerous spurious FPS actuations initiated by operator testing errors or by maintenance activities, e.g., welding, and steam or high humidity in the vicinity of FPS detectors or control circuitry and components. An NRC memorandum (Ref.2), issued on January 28, 1982, provided additional examples of FPS actuation interactions and suggested that all types of FPS, e.g., water, halon, carbon dioxide and other chemicals be included in a review of their safety significance and possible corrective steps.

FPS actuations that result in adverse interactions with plant systems needed to achieve safe plant shutdown or to mitigate a postulated accident reduce the availability of such systems. This concern is accentuated when common cause initiators and common mode failures of safety-related equipment are considered. Examples of common cause initiators include earthquakes, smoke and heat intrusion into multiple fire zones, and fire suppressant intrusion into multiple fire zones affecting several safety-related systems. Examples of common mode failures of safety-related systems and/or auxiliary systems supporting safety-related systems include electrical shorts in instrument cabinets and electrical power distribution centers, CO_2 ingress into the fresh air intake of emergency diesel-generator sets, and cold CO_2 induced thermal stresses and cracking of station battery casings, with loss of offsite power during an earthquake. It should be noted that a number of common cause initiators and common mode failures are not mutually exclusive and they may be part of a single event sequence.

Four plants representative of the various designs of currently operating nuclear power plants were evaluated as part of this issue (Refs. 4–8). Furthermore, a generic evaluation of this issue (Ref. 9) was performed taking into account the insights from the technical findings of these four evaluations, as well as design and plant layout information of a large number of operating plants collected for this purpose. An extensive review of operational experience involving actuations of fire protection systems was performed prior to the analytical assessments of risk associated with this issue. The review of the operational experience and the aforementioned analyses showed the following:

- 0.15 inadvertent FPS actuations/RY
- 0.02 advertent FPS actuations/RY
- 37% of all actuations damaged some equipment

- 20% of all actuations resulted in a plant transient and reactor trip

- Core Damage Frequency (CDF) contributions from GI–57 root causes for the four individual plants evaluated were estimated to be in the range of 7.3E–06/RY to 5.6E– 05/RY.

- Dominant risk contributors are associated with seismic/FPS and seismic/fire interactions resulting in sequences involving station blackout and small LOCAs

Both of these categories of dominant sequences are currently being addressed by the Individual Plant Examination of External Events (IPEEE) program. Generic Letter 88–20, Supplement 4, June 27, 1991 (which initiated the IPEEE) states:

"The walkdown procedures should be specifically tailored to assess the remaining issues identified in the Fire Risk Scoping Study: (1) seismic/fire interactions, (2) effects of fire suppressants on safety equipment, ..." (paragraph 4.2).

The staff notes that any complete assessment of seismic/fire interactions would necessarily include seismic effects on manual firefighting, since automatic fire suppressant systems are not seismically qualified, thus increasing the potential need for effective manual firefighting during seismic events.

NUREG–1407 ("Procedural and Submittal Guidance for the IPEEE for Severe Accident Vulnerabilities", June, 1991) reiterates the above (in the first paragraph of section 4) and also adds:

"The use of an existing fire PRA for the internal fires IPEEE is acceptable provided the PRA reflects the current as-built and as-operated status of the plant and the licensee addresses the deficiencies of past PRAs that are identified in the Fire Risk Scoping Study (NUREG/CR–5088). Deficiencies may include the use of low conditional failure probabilities for dampers and penetrations, no consideration of damage from the use of fire suppressants, inappropriate estimates of the effectiveness of manual fire fighting, and no consideration of seismic/fire interactions." (paragraph 4.2).

Also in NUREG–1407, Appendix D, section 6 ("Internal Fires") the staff response to question 6.2 states:

"The procedurally directed walk-downs associated with internal fires vulnerability evaluation can be planned as part of the seismic walk-downs that would specifically look for the seismic-induced fire vulnerability issues. The idea is to first identify those areas that could be vulnerable so that they can be brought into focus during the walkdown.

> "For example, if a plant didn't have its diesel fuel tank strapped down properly one could postulate a large fuel source for fire as a result of a seismic event. Other similar seismic/fire interactions were summarized in Section 7 of NUREG/CR-5088."

In addition, for performance of the IPEEE, many licensees will also be using the Fire-induced Vulnerability Evaluation (FIVE) methodology developed by the Electric Power Research Institute (EPRI) as described in "Fire-induced Vulnerability Evaluation (FIVE)", EPRI TR-100370, April, 1992. Following a description of the three basic Phases of FIVE, that document states:

> In addition, there is a discussion of several potentially risk significant items that were identified in the NUREG/CR-5088, "Fire Risk Scoping Study", that **should also be considered in performing FIVE.** (Section 7.0) [**emphasis** added].

Section 7.0 of NUREG/CR-5088 is a description of the Sandia Fire Risk Scoping Study Evaluation which includes discussion of the dominant risk contributors to GSI-57-related events, including Seismic/Fire interactions (seismically induced fires, seismic actuation of fire suppression systems, and seismic degradation of fire suppressant systems), and Manual Fire Fighting Effectiveness. The following seismic/fire interactions are given as examples of the situations to be considered by the IPEEE: unanchored CO_2 or Halon tanks, possible relay chatter in fire protection system actuation systems, fire alarm systems having only a smoke-actuated alarm without a heat or flame detector, fire pump mounts without vibration amplitude stops, cast iron fire mains, inadequate anchoring of electrical cabinets and inadequate slack in the wires leading to such cabinets (to avoid sparks from tight wires), unanchored high-pressure gas bottles, potential interactions between sprinkler system heads and adjacent pipes, and presence of mercury switches in fire suppression and detection systems (such switches should be replaced with alternate, seismically insensitive switches).

For the "generic" evaluation, after subtraction of CDF contributions from GSI-57-related events involving:

(a) Seismic/Fire: – seismic induced fire plus seismic induced suppressant diversion. The unsup-

pressed fire and/or the diverted suppressant incapacitate safety related equipment needed to mitigate effects of the seismic event; and

(b) Seismic/FPS: – seismic induced actuation of the FPS. Released suppressant damages safety-related equipment needed to mitigate effects of the seismic event,

which are being emphasized by the IPEEE, the mean CDF of the remaining contributors is less than 1.0E–05/RY which does not justify a generic backfit.

The risk reduction estimates, cost/benefit analyses, and other insights gained during this effort have shown that implementation of the recommendations contained in this report can significantly reduce risk, and that these improvements can be warranted in accordance with the backfit rule, 10 CFR 50.109(a)(3). However, plant specific analyses are required in order to identify such improvements. Generic analyses can not serve to identify improvements that could be warranted for individual, specific plants. Plant specific analyses of the type needed for this purpose are underway as part of the Individual Plant Examination of External Events (IPEEE) program.

Hence, we concluded that GSI-57 should be resolved with no new regulatory requirements generically imposed for existing plants.

For new plants, the enhanced safety requirements given in SECY-90-016, "Evolutionary Light Water Reactor (LWR) Certification Issues and their Relationship to Current Regulatory Requirements", January 12, 1990, allow the same GSI-57 resolution to be applied (i.e., with no new regulatory requirements imposed). This is justified since the designers of new plants will perform plant-specific PRAs, including external events, that are equivalent to the IPEs and IPEEEs which will resolve this issue for existing plants.

However, the reports dealing with the evaluation of this issue, including this regulatory analysis, will be published, or otherwise be made publicly available, so that the insights gained from this effort may be used by licensees and applicants, and the nuclear industry in general, to take voluntary steps for plant-specific cases, including frontfitting considerations for advanced or evolutionary reactor designs.

7 IMPLEMENTATION

We concluded that GSI-57 should be resolved with no new regulatory requirements generically imposed for existing plants, and that cost-effective modifications that may be desirable on specific plants will be identified by the IPEEE program.

It is expected that the "design-specific probabilistic risk assessment" (PRA) required for new plants by 10CFR52.47.(a).(v) will be equivalent to the IPEs and IPEEEs being performed for existing plants, i.e. it will consider external events including the situations which contribute significantly to CDF due to GI-57 related events, and it will identify plant specific modifications that may be desirable.

We therefore concluded that GSI-57 should be resolved with no new regulatory requirements generically imposed for new plants, and that cost-effective modifications that may be desirable on specific new designs will be identified by the PRAs that will be performed for those designs.

8 REFERENCES

1. Memorandum from G. Lainas (NRC:NRR) to D. Eisenhut (NRC:NRR),"Summary of the Operating Reactor Events Meeting on January 7,1982," January 13, 1982.

2. Memorandum from C. Michelson (NRC:AEOD) to R. Vollmer (NRC:NRR) and E. Jordan (NRC:IE), "Effects of Fire Protection System Actuation on Safety-Related Equipment," January 28,1982.

3. IE Information Notice 83-41, "Actuation of Fire Suppression System Causing Inoperability of Safety-Related Equipment," U. S. Nuclear Regulatory Commission, June 22, 1983.

4. J.A. Lambright, et al., "Risk Evaluation for a Westinghouse PWR, 'Effects of Fire Protection Systems Actuation on Safety-Related Equipment' (Evaluation of Generic Issue 57)," NUREG/CR-5789, SAND91- 1534, December 1992.

5. J.A. Lambright, et al., "Risk Evaluation for a General Electric BWR, 'Effects of Fire Protection Systems Actuation on Safety-Related Equipment' (Evaluation of Generic Issue 57)," NUREG/CR-5791, SAND91- 1536, December 1992.

6. J.A. Lambright, et al.,"Risk Evaluation for a Babcock and Wilcox PWR, 'Effects of Fire Protection Systems Actuation on Safety-Related Equipment' (Evaluation of Generic Issue 57)," NUREG/CR-5790, SAND91- 1535, September 1992.

7. G. Simion, et al.,"Risk Evaluation of a Westinghouse 4-Loop PWR, 'Effects of Fire Protection System Actuation on Safety-Related Equipment (Evaluation of Generic Issue 57)'," EGG-NTA-9081 Letter Report, Idaho National Engineering Laboratory, December 1991.

8. Letter Report, "Seismic Risk Evaluation for a Water Reactor, 'Effects of Fire Protection System Actuation Safety-Related Equipment'," Sandia National Laboratories, December 1991.

9. J. A. Lambright, et al.,"Risk Evaluation of Generic Issue 'Effects of Fire Protection System Actuation on Safety-Related-Equipment'," NUREG/CR-5580, SAND90-1507, December 1992.

10. Memorandum from Edward L. Jordan (CRGR: NRC) to Eric S. Beckjord (RES:NRC), "Implementation of the Safety Goals," September 1990.

11. Electric Power Research Institute, Report on the Effects of Fire Suppressants on Electrical Components in Nuclear Power Plants, Report 01-0650- 1737, Revision C, Impell Corp., November, 1989.

12. R. B. Worrell, "SETS Reference Manual," NUREG/CR-4213, S and 83-2675, May 1985.

13. W. T. Wheelis, "User's Guide for a Personal-Computer-Based Nuclear Power Plant Fire Data Base," NUREG/CR-4586, SAND86-0300, Sandia National Laboratories, 1986.

14. Top Event Matrix Analysis Code, NUREG/CR-4598, Sandia National Laboratories, January 1989.

15. Latin Hypercube Sampling Code, NUREG/CR-3624, Sandia National Laboratories, March 1984.

16. Heaberlin, S., et al., "A Handbook for Value/Impact Assessment," NUREG/CR-3568, December, 1983.

17. Claiborne, E., et al., "Generic Cost Estimates," NUREG/CR-4627, Revision 1, Science and Engineering Associates, Inc., February, 1989.

18. Lopez-Vitale, B., et al., FORECAST 3.0 User Manual, SEA Report No. 89-461-04- A:1, April 1, 1990.

19. The Energy Economic Data Base (EEDB)

    a. "Phase VIII Update (1986) Report for the Energy Economic Data Base Program," DOE/NE-0051/1, United Engineers and Constructors, August, 1986.

    b. "Phase VIII Update (1986) BWR Supplement for the Energy Economic Data Base Program," NUREG/CR-4764, United Engineers and Constructors, December, 1986.

20. "1990 Edition of Mechanical and Electrical Cost Data," Kingston, MA, R. S. Means Company, Inc., Annual, 1990.

21. Claiborne, E. et al., "Cost Analysis for Potential BWR Mark I Containment Improvements," NUREG/CR-5278, Science and Engineering Associates, Inc., January, 1989.

22. Memorandum from E. S. Beckjord, RES Office Letter No. 2, "Procedures for Obtaining Regulatory Impact Analysis Review and Support," November 18, 1988.

23. Memorandum from E. S. Beckjord to Distribution, RES Office Letter No. 3, "Procedure and Guidance for the Resolution of Generic Issues," May 10, 1988.

NRC FORM 335
(2-89)
NRCM 1102,
3201, 3202

U.S. NUCLEAR REGULATORY COMMISSION

BIBLIOGRAPHIC DATA SHEET

(See Instructions on the reverse)

1. REPORT NUMBER
(Assigned by NRC, Add Vol.,
Supp., Rev., and Addendum Numbers, If any.)

NUREG–1472

2. TITLE AND SUBTITLE

Regulatory Analysis for the Resolution of Generic Issue 57: Effects of Fire Protection System Actuation on Safety-Related Equipment

3. DATE REPORT PUBLISHED

MONTH	YEAR
October	1993

4. FIN OR GRANT NUMBER

5. AUTHOR(S)

H. W. Woods

6. TYPE OF REPORT

Regulatory

7. PERIOD COVERED (Inclusive Dates)

8. PERFORMING ORGANIZATION – NAME AND ADDRESS (If NRC, provide Division, Office or Region, U.S. Nuclear Regulatory Commission, and mailing address; If contractor, provide name and mailing address.)

Division of Safety Issues Resolution
Office of Nuclear Regulatory Research
U.S. Nuclear Regulatory Commission
Washington, DC 20555–0001

9. SPONSORING ORGANIZATION – NAME AND ADDRESS (If NRC, type "Same as above"; If contractor, provide NRC Division, Office or Region, U.S. Nuclear Regulatory Commission, and mailing address.)

Same as above.

10. SUPPLEMENTARY NOTES

11. ABSTRACT (200 words or less)

Actuation of Fire Protection Systems (FPS) in Nuclear Power Plants have resulted in adverse interactions with equipment important to safety. Precursor operational experience has shown that 37% of all FPS actuations damaged some equipment, and 20% of all FPS actuations have resulted in a plant transient and reactor trip. On an average, 0.17 FPS actuations per reactor year have been experienced in nuclear power plants in this country. This report presents the regulatory analysis for GI–57, "Effects of Fire Protection System Actuation on Safety-Related Equipment". The risk reduction estimates, cost/benefit analyses, and other insights gained during this effort have shown that implementation of the recommendations contained in this report can significantly reduce risk, and that these improvements can be warranted in accordance with the backfit rule, 10 CFR 50.109(a)(3). However, plant specific analyses are required in order to identify such improvements. Generic analyses can not serve to identify improvements that could be warranted for individual, specific plants. Plant specific analyses of the type needed for this purpose are underway as part of the Individual Plant Examination of External Events (IPEEE) program.

12. KEY WORDS/DESCRIPTORS (List words or phrases that will assist researchers in locating the report.)

generic issue
fire protection system
regulatory analysis
safety-related equipment

13. AVAILABILITY STATEMENT
Unlimited

14. SECURITY CLASSIFICATION

(This Page)
Unclassified

(This Report)
Unclassified

15. NUMBER OF PAGES

16. PRICE